Gardening On Sandy Soil

IN NORTH TEMPERATE AREAS

by Christine Kelway

WITH A NEW FOREWORD BY

Daniel J. Foley

DOVER PUBLICATIONS, INC.
New York

This Dover edition, first published in 1975, is an unabridged republication of the work originally published with the title *Gardening on Sand* by W. H. & L. Collingridge Ltd., London, in 1965. The present edition is published by special arrangement with David & Charles Ltd., Newton Abbot, Devon, England. A new Foreword has been written specially for the Dover edition by Daniel J. Foley.

International Standard Book Number: 0-486-23199-2
Library of Congress Catalog Card Number: 75-13386

Manufactured in the United States of America
Dover Publications, Inc.
180 Varick Street
New York, N.Y. 10014

Foreword

to the Dover Edition

THE study of soils is often a bewildering subject for gardeners, especially for those accustomed to a site where a fair amount of garden loam exists. When challenged to make a garden on sand or on a sandy soil containing a minimum of organic matter, the beginner often finds the task insurmountable. Even the experienced amateur may at first be frustrated. Actually, no type of soil, sandy or clayey, is lacking in possibilities for a rewarding garden, if it is improved adequately. All it requires is know-how, time and labor. Christine Kelway, an accomplished horticulturist of solid background and wide experience, approaches the subject of sand gardening with sound common sense as well as a keen understanding of the problems of the amateur gardener. True, she gardens in the mild climate of England where winters are not usually severe, but the techniques she employs and the principles she advocates are similarly applicable to most sandy soils in the United States.

Some of the materials she uses are not readily available here, but there are adequate substitutes. For example, she recommends bracken (the stems and fronds of the royal fern, *Osmunda regalis*), which is common in England, for making compost and for use as mulch to build up calcareous sand. Most American gardeners would find it difficult to obtain bracken in sufficient quantity to secure the desired results, despite the fact that the royal fern is one of our native species. However, composted bark or composted wood chips (both of which are now widely used for mulching gardens), as well as seaweed, peat moss, leaf mold and compost can be incorporated with the sand to build up the needed organic matter or humus content which sandy soils require. These materials retain the needed moisture and plant nutrients (supplied by fertilizer) which growing plants require to produce healthy growth.

Acidifying soil agents such as aluminum sulphate or dusting sulphur will help materially to create the required acidity level for those plants which need it. The specific amounts to be used in a given area are most wisely determined by having the soil tested. Also, several commercially prepared materials, sold under various trade names, are readily available to improve soil acidity. Instructions printed on the container need to be followed precisely.

Home gardeners are rapidly learning the value of soil testing. While there are test kits available that are simple and easy to use, they are not always completely satisfactory. The beginner—and for that matter the gardener with a fair amount of experience—may obtain entirely satisfactory results by having tests made by competent soil analysts at his county or state experiment station. This service is usually free. A telephone query or a letter will result in a set of instructions to be followed in preparing and sending the test soil sample.

The author places prime emphasis on the importance of mulching the sandy garden to conserve moisture and control weeds. This practice is essential to the success of the sandy garden, in various parts of the United States, particularly during seasons of drought. In addition to the mulches mentioned, the following suggestions are offered as a supplement.

Bark, chopped or shredded. Obtained from sawmills, it is readily available at garden centers and nurseries. The bark derived from hemlock, pine, spruce and other trees, often with a little birch mixed in, is shredded and composted for several years before it is sold. It makes an excellent mulch and may be used to advantage throughout the garden. Unlike redwood bark, this composted material breaks down rapidly and needs replenishing every two years. It is also an excellent and efficient source of organic material for incorporating with sand to build up the essential organic content.

Cranberry vines, sometimes offered commercially as Cran-mulch, are used for winter mulch in areas to which they can be shipped at reasonable cost. The material, handled in bales, is easy to spread, making a most attractive and effective form of winter cover for perennials and shrubs. It is also an ideal mulch for natural plantings and wild gardens.

Peat moss is easy to handle as a mulch; cover should be at least an inch deep. The coarse brown peat sold in bales (usually imported), derived from sphagnum moss, makes a satisfactory cover. However, it tends to cake when the top surface dries, preventing water from penetrating easily. This condition may be overcome by mixing it with soil, pine needles or chopped bark. Peat from local bogs is often more desirable, since it is finer in texture and often can be obtained at moderate price if purchased in quantity. Actually, the finer grades of peat are easier to spread than the sphagnum type (which is fibrous in texture), and they are also easier to soak. Peat absorbs ten times its weight in water, and this point should be borne in mind when applying it. Never spread bone-dry peat on soil as a mulch or incorporate it with fairly dry soil to improve texture without moistening it thoroughly. Dry peat tends to pull the moisture from the soil with which it comes in contact.

Redwood bark is sold in bags in most garden centers. Because of its coarse texture (graded from coarse to fine), this material is not blown away easily by wind; it allows water to penetrate the soil easily, breaks down slowly, and is dark in color. (Coconut and yucca fiber are also considered useful mulches on the West Coast.)

Salt marsh hay is primarily a winter mulch, spread after the ground has frozen to protect perennials, rock plants and roses. It may be used to good advantage for strawberries and vegetables, or among shrubs. As it rots, it provides organic matter for the soil. However, despite its fine texture as contrasted with straw, it is not particularly pleasing to the eye in tailored plantings of perennials and annuals. For winter protection, it is ideal and easy to handle. In spring it can be raked and used for the purposes suggested as a summer mulch, or it can be added to the compost pile to produce humus.

Wood chips make an excellent mulch for more permanent use, because they are slow to deteriorate. The particles of various common hard and softwood trees vary in size, allowing water to penetrate the soil easily. Although light in color when applied, this material weathers in appearance as it ages, and its rough texture is pleasing to the eye. Apply it at least two inches deep. When new, it can be darkened by scattering a thin cover of peat, pine needles or soil over it. Composted wood chips are available from some garden centers and nurseries. Arborists and municipal tree departments are excellent sources, since practically all of them use chipping machines to reduce the bulk of cut branches. Leaves and twigs make the mixture coarse at first, but it soon breaks down. Truckloads are often available at small cost. Sprinkling the mulch with a complete fertilizer will compensate for the nitrogen extracted from the soil as the wood particles break down.

Washed gravel, in several sizes varying from one-eighth inch (referred to as peastone) up to three-quarters of an inch in diameter, makes an excellent permanent mulch for those who garden on sand. It needs to be spread at least one inch deep to be effective in conserving moisture and stifling weeds. Even so, some weeds will appear from time to time, but they are easily controlled with a stone mulch. It is pleasing to the eye, natural in color and ideally suited to the sand or seaside garden. The labor cost involved in spreading the stone is largely offset by the moderate cost of the stone itself.

As Christine Kelway so aptly declares, "the keynote to success in a garden is to put the right plant in the right place, i.e. to choose a plant which will reach the height of its beauty in the soil and climate provided for it." In describing the shrubs, perennials, trees, bulbs and annuals specially suited and adapted to sandy soils, she includes, along with the hardy old standbys, many subtropical kinds that flourish in the mild climate of England. American gardeners perusing these lists will soon realize that the natives of Australia, New Zealand, Mexico, California and South America are suited only to similar climates south of Washington and along the West Coast. Gardeners in the northeastern United States must, for the most part, be content with only the hardiest kinds. Yet some of these so-called tender plants, as we refer to them, can be enjoyed during the summer months if wintered over in a cool greenhouse, a pit or a storage cellar. Considered as a whole, the entire list of "dry-soil" plants is as impressive as it is comprehensive. At best small gardens can hold but a few. The fun is to be able to pick and choose.

Similarly, for many of the roses and rhododendrons, heaths, heathers, brooms, and other kinds, substitute varieties will need to be chosen by American gardeners, since not all those listed are readily available here. As with any highly specialized form of gardening dictated by specific soil conditions, certain plant materials need to be obtained from specialists. Most nurseries and garden centers are able to supply the more commonly planted kinds. Growing ornamental plants on sand can be not only a highly rewarding adventure but one rich in the sense of achieving results under what may at first glance appear to be adverse conditions.

DANIEL J. FOLEY

Salem, Massachusetts
April 4, 1975

Gardening
On Sandy Soil

Contents

Illustrations

Acknowledgements

I am most grateful to Mr John Street for writing the chapter 'Rhododendrons on Sand', and to Mr Rowland Jackman, Mr H. J. Randall, Mr N. Treseder and Mr Graham Thomas for their helpful advice. I would also like to thank the following for supplying photographs: Messrs. J. McBain Allan, F. J. Chapple, P. Hunt, A. J. Huxley, R. V. G. Rundle and H. Smith.

Introduction

Seldom are we in the enviable position of being able to choose the kind of soil on which we garden, for circumstances beyond our control are often the deciding factor; so that a good fat loam remains the gardener's dream and we are left with either a heavy clay or a hungry sand.

One way to overcome such difficulties is to learn to live with them and this is nowhere more true than in our gardens, where a garden on sand presents its own problems.

Very broadly speaking, sandy soils occur where solid sandstone formations or superficial spreads of sand are found, and sand is more widespread in the British Isles than is generally supposed. It is not possible to obtain a clear picture of those districts where sand exists owing to its wide and irregular distribution, but some of the better known areas are the strip known geologically as Bagshot Beds, from Bagshot in Surrey through Hampshire to the New Forest and into Dorset, the Brecklands in East Anglia and the Woburn Sands in Bedfordshire. These are just a few outstanding examples of where sand lies. There are also extensive belts of sea-sand around our coasts which differ in having a high lime content, distinct from the acid sand of other areas.

Lack of plant foods, lack of organic material to give bulk and body, and lack of moisture where the water-table is low, are some of the main problems of the gardener on sand, while the shifting of sandy soils and wind erosion can be very troublesome, particularly on the coast.

Although the fixing of coastal sands and their afforestation is beyond the scope of this book, there are gardeners who include some sand-dunes within their boundaries, and these and other matters and the plants most suited to sandy soils are dealt with in the pages which follow.

C. K.

Trebetherick,
  Cornwall.

CHAPTER ONE

Soils

IN his *English Flower Garden* William Robinson wrote that, next
to position, soil is the most important element in the formation of a
garden, and we have only to study the natural vegetation around us to
realise that the behaviour of plants is largely determined by the soil
they live in.

In recent years our scientific knowledge of the soil has increased
enormously, and this has a definite bearing on the kind of garden we
can make and which plants will grow in it. Yet I am continually being
surprised by the number of people who neither know nor care what kind
of soil they are taking on when building or buying a new home, these
same people only living to regret it when it turns out to be a true clay,
heavy to cultivate and unworkable for many months of the year.

What then are the different types of soil so vitally important to
gardeners? Since the ingredients of Mother Earth are sand, clay and
humus, the greater or lesser degree in which each is found decides the
type of soil. The ideal for garden-making would be a deep loam, free
from all calcareous substances; but nothing is ever quite what it should
be, and only the fortunate few find themselves in a Garden of Eden
without the serpent. The wisest course is to seek to improve the soil
that circumstances have provided, growing in it a preponderance of
those plants which by nature are best suited to it.

Sand and Clay. The ordinary gardener thinks in terms of 'light' or
'heavy' soil; a light soil, at one end of the scale, is almost pure sand,
and at the opposite extreme is the heavy one, which is clay, though
strangely enough, taken bulk for bulk, sandy soils are the heaviest
of all soils. Neither makes for good gardening in its natural state,
though were I forced to choose, it would be sand every time.

A dictionary description of 'sand' reads: 'Fine particles of crushed
or worn rocks containing no humus or earthy matter'. A sandy soil
contains a high percentage of sand; the experts give it as not less than
60 per cent., and the notoriously infertile Bagshot Sands vary from 40
to 70 per cent. Sand can be fine, or coarse, as in the case of gravel.

Gravel has all the characteristics of a coarse sand and is particularly unfavourable to plant life. Because of its poverty in nutrients and because of its arid dryness, excessive rain, which might be thought to be beneficial on this type of gravelly soil, often leads to severe leaching.

Being largely composed of relatively big, rough particles, and the air spaces between them being bigger than those on a heavy soil, sand holds air and therefore warmth, but not water. On clay quite the opposite is true: the particles are exceedingly small and pack down so tightly that at times the clay becomes almost impervious; in wet weather it becomes unbearably sticky, while during drought the ground bakes into a hard, solid mass and with some clays cracks into deep ruts. While clay provides plant food and has fertility, pure sand has no food value in itself and its properties are largely physical and mechanical. It is as warm as clay is cold, and whereas clay is stiff or sticky on the spade, sand falls readily into such fine particles that it slips too easily through a gardening fork. Though soil texture can be determined most accurately in the laboratory, the experienced gardener can tell it by its feel. Sand feels gritty, but clay is sticky when wet and harsh when dry, though only an experienced gardener can estimate the amount of each by rubbing a pinch between his fingers.

Humus. The third ingredient of Mother Earth is humus—the result of disintegration and decay of organic matter over the years; it is mainly derived from dead roots and a surface litter of dead leaves and shoots as well as dead worms and insects. Soil bacteria ultimately reduce all this material to simple substances that are available to plants. Humus can absorb and retain a great deal of moisture and it has a dark colour, either black or brown. On woodland soils the layer below the surface litter is rich in humus and invaluable for incorporating into light sands. In soils most favourable to plant growth the decay of humus is rapid but in opposite soil types such as sand, the decay is very slow indeed. Unfortunately, on excessively poverty-stricken land the amount of vegetation is nothing like as abundant as on richer, heavier land; the resulting humus is therefore correspondingly less, and since worms and insects are found in greater numbers where there is a good supply of humus, it is a vicious circle which will continue unless humus is brought in from outside.

Soil is far from being an inert mass; it is teeming with multitudes of living organisms which make their home in it and play an important part. One of the most valuable is the earthworm.

Earthworms. Worms, we know, play a large part in helping to build up the structure of the soil; they drag down decayed material into their holes and I am always delighted to see them at work in our sandy garden, as this tells me they can find something of value to take down to their earthy lairs. You have only to turn over a heap of 'muck' to see that this is the stuff for them. Darwin wrote: 'Long before man existed, the land was, in fact, regularly ploughed and still continues to be ploughed, by earthworms.' He estimated that in an average acre of land there were about 53,000 worms, and later this was found to be an underestimate. Everything in one's power should be done to build up the worm population and it is unfortunate that farmyard manure, which is so rich in worms, is practically unobtainable by many gardeners.

Worms are rare on dry, sandy soils and heathland, because they perish if deprived of moisture. They tolerate a certain amount of acidity in the sand, though there are practically none found below a pH of 4·5. Where acidity is extreme they do not survive, so that dead vegetation accumulates on the surface, ultimately to become peat. Years ago, when we started to make our garden on the calcareous sand of North Cornwall, the worm population was negligible; now, after the incorporation of pig and cow manure, loads of compost and soft green material, there are few parts of the garden where worms cannot be found.

I know that many people are more concerned with their destruction than in increasing their numbers, and it is true they disfigure grass lawns, but this damage is somewhat overrated, particularly on sandy soils where it is only necessary to brush the worm-casts with a besom to get rid of them. Moreover, this nuisance is more likely to occur on low-lying, badly drained ground than on sand. In the soil, worms are the gardener's best friend, aerating it, dragging down leaves and compost into it, passing it through their bodies, mixing it with lime from their glands, and even in death forming a useful part of the soil's composition.
Two kinds of sand. That there are two kinds of sand, differing greatly in character, is not always recognised sufficiently by gardeners. Acid sand and alkaline sand are the two kinds with which gardeners must concern themselves. The greatest stretches of sand in this country are extremely acid; these are found most often in inland districts, while limy or alkaline sand lies chiefly in pockets on the coast. In inland areas coarse acid sand may lie to a depth of a hundred or more feet, though it is often covered by a thin skin of black, peaty earth, which overlies the yellow sand. In places this may not be more than an inch thick, though in hollows where humus has collected it may be as much

as 4 inches or more; this is the result of the accumulation of decayed material. It is valuable as plant food and permits a different vegetation to that found on fine sea sand. Here vegetation is so scanty, owing to strong gales and salt, that there is often nothing but blown sand, rich in lime but poor in humus.

Effect of rainfall. Rainfall largely determines the properties of a sandy soil. High rainfall washes out the nitrates so easily that plants fail to find sufficient nitrogen for their needs. It also leaches out much of the calcium and magnesium, so that the soil becomes acid; and, at a further stage, acidity becomes more pronounced as much of the iron, aluminium and manganese is leached from the surface soil. The result is often three layers. First, the topsoil, black or brown and strongly acid, consisting mainly of plant material in process of decomposition. Second, the leached layer of sand, which, having lost its iron and manganese, is pale in colour and desperately poor and unrewarding for the cultivator. Third, a layer containing these bases and aluminium, which have been precipitated here, giving the soil a brown colour. This layer, in which iron has been deposited, often settles into a hard pan, hampering root development. Underneath again may be the original rock material which remains unchanged.

Even sandy soils uncultivated by man are rarely untouched; their herbage is grazed either by sheep or rabbits, and where grazing is light, spreads of heather (*Calluna*) and heath (*Erica*) occur, but on more heavily grazed land there is grass. Conifers flourish on acid sand, tolerating both poverty of soil and acidity, and once past the nursery stage they grow satisfactorily. Large areas of such land have been planted by the Forestry Commission.

Not only on sea sand but also on moors and marshes soil will be found far too sandy to be fertile. Marsh sands will be mixed with a good deal of humus, moor sands may be covered with a coating of rough peat, while coastal sands are little but sand.

That sand is basically infertile, lacking in humus, devoid of worms and without the water-holding capacity of clay, cannot be disputed, yet it can be improved out of all recognition, and with greater ease than its opposite number, the true clay soil. The late V. Sackville-West, who gardened on each in turn, described clay as 'dire, plaguey, baleful, thankless soil—workable for about five days out of the year', and the best advice she could give was 'to remove your dwelling as speedily as possible to another place'.

CHAPTER TWO

The Pros and Cons of Sandy Soil

Few advantages in life are achieved without some compensating disadvantage or drawback; what we gain in one direction we often lose in another, and it is not always easy to assess the true value of factors on either side. I shall try, however, to set out both the limitations and advantages of gardening on poor, hungry, sandy soil as I see them.

LIMITATIONS

However much soils may differ in character and productivity at the outset, we can take comfort in the fact that few soils are hopeless or beyond redemption, though if we intend to 'make the desert blossom as the rose', we should first face up to the difficulties. Poor, thin sand has very definite drawbacks and one of the most obvious is its lack of humus and, therefore, lack of fertility.

Lack of humus. Because sand, unlike clay, contains no plant food, it is greedy for anything it can absorb. Try as you will, you will not grow first-class flowers or vegetables on excessively sandy soil unless you feed it, not only at the start, but continuously. When we made our garden some years ago, there was a small patch of extremely sandy soil which we hoped to make into a vegetable garden. The soil was almost pure sand, very light and yellow, not coarse, inland stuff, but fine sand blown by the wind from the neighbouring beach, so fine that it slipped through the fingers; it was also highly alkaline, containing a large proportion of crushed sea-shells. With little knowledge then of sand, and of limy sand in particular, we sowed some ordinary garden peas in this unpromising bit of ground. They germinated quickly in the warm sand but they never grew higher than a couple of inches—there simply was not sufficient nourishment for them. Over the years, this same intensely sandy piece of ground has been extensively fed with strawy farmyard manure, garden compost and any soft green material we could put in. Everything possible was done to increase its fertility, and today it grows excellent crops of vegetables.

Additions to the soil, however, such as manures, peat, or spent hops, are not cheap, and they entail a great deal of labour. And it must be carried out regularly if plants are to continue to flourish, for there is no letting up with sand until it has lost its pale look and has begun to darken.

Lack of moisture. As well as being a 'hungry' soil, sand is also a 'thirsty' one, and since it is composed of comparatively large, rough particles which hold air but not water, it is very porous. It dries out rapidly and cannot retain moisture, so that though we are often told that hot summers are a thing of the past, only a day or two without rain turns the garden on sand into a dusty waste. We have a saying in North Cornwall, where the rainfall is lower than in the southern half of the county, that we need a shower of rain every day and two on Sunday. If short, rainless periods have some effect, prolonged drought works havoc among plants on dry, sandy soil. The flowering season of plants is seriously reduced and perennial plants which in normal conditions should continue in flower from four to six weeks, will be past their best in half that time.

Indeed, lack of moisture can fairly be considered one of the most dangerous enemies of plants, for water is vital to plant growth and it is doubtful whether it is possible to over-water a healthy plant in the full vigour of its growth, so long as the water is able to drain away.

The nearest approach to drought in this country is experienced in those districts where sand or chalk predominate, particularly where the sand is extremely alkaline. Coarse sand or gravel has less water-retaining capacity than fine sand, and where the sand area lies high, it is dependent on its rainfall and is very difficult to deal with, since there is no water draining down from a higher level. Such areas are often excessively arid, and hot, southerly slopes are also a problem challenging the ingenuity of the gardener. Here it is absolutely essential to provide the soil with those ingredients which alone enable it to retain moisture long enough for the plants to avail themselves of it. These sandy hill-top gardens are among the most heart-breaking there are. The ingredients for providing humus, in the form of manures, material from the compost heap, or peat, must all be wheeled up these sloping gardens, and annual mulches, without which these dry soils will lack the bacteria necessary for breaking down the chemicals into plant food, are almost impossible to apply—they keep slipping down to a lower level.

Indeed, those who are considering making a garden in such conditions should ponder on the difficulties.

1 *Above:* The author's Cornish garden on sand. Spanish gorse is in the foreground, and broom to the right of the picture

2 *Ceanothus thyrsiflorus* is one of the toughest and hardiest of the evergreen ceanothus. It has bright blue flowers in early summer

3 *Above: Cistus* Paladin Pat has white flowers, 5 inches across, blotched with crimson

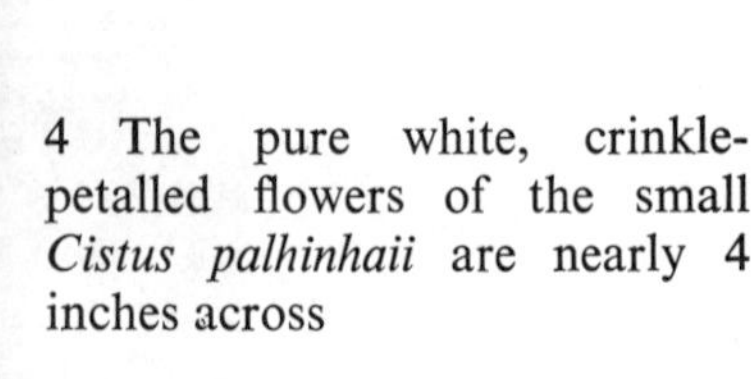

4 The pure white, crinkle-petalled flowers of the small *Cistus palhinhaii* are nearly 4 inches across

5 A yellow-fruited cotoneaster can be most attractive against a wall or fence, and provides a change from the usual red-fruited species and varieties

Loss of moisture caused by spring winds. Few things are so damaging to plants or so dismaying to the gardener's heart as those cold, drying winds of spring. Though they help to dry out a heavy clay soil after the winter rains, they do great harm on sand unless plants are kept well watered. Roots are quite unable to replace the loss of moisture caused by the rapid evaporation on the leaf surfaces, while sharp drainage and cold winds combine to turn a garden on sand into a semi-desert.

Newly-planted subjects are the first to suffer, needing special care and attention until they become established. For this reason spring-sown annuals are rarely as fine as those sown in autumn, unless the spring is an unusually wet one.

Underground moisture. We may as well face it, a garden on sandy soil is a thirsty garden, and it is important at all times to conserve and use the moisture in the subsoil. Underground water has a definite bearing on the fertility of light soils. The Weybridge soil in Surrey, for instance, has an underground source of water only 3 feet below its surface and is excellent for growing wheat, whereas the notoriously infertile Bagshot Sands and the barren Hythe beds in West Surrey, though similar in physical type, have in general no such water supplies and are sterile because they lack the clay which alone could confer an adequate power of water retention.

Coolness of climate, as well as underground sources of water, has also an effect on soils containing a high percentage of sand. Soils containing so much sand that they burn and scorch in a dry, hot district, prove very suitable for cultivation in a cooler district where lower temperatures prevail. For example, potatoes will grow very well indeed on light soils which are cool and moist.

The behaviour of sandy soils towards vegetation depends largely on their position, their depth and the nature of the subsoil. Sandy soil over a bed of gravel or a layer of rock, e.g. a hard pan, is likely to be either parched or waterlogged, and its water supply in either case will be so unsatisfactory that cultivation is unprofitable.

Watering the sandy garden. The garden on sand requires constant watering, for even a desert soil with scanty vegetation and little organic matter has been known to be made fertile when irrigated. Simple as watering may seem, there is a right and wrong way of doing it. How tempting it is after a hot day when the ground is parched and dust-dry, to wander round the garden, can in hand, lightly spraying each and every wilting plant, a bit here and a bit there.

One of the greatest mistakes we can make, however, is to add a little dribble of water which merely moistens the surface, leaving the root-run dry. It encourages the roots to come up in search of moisture, when the poor things get scorched and burnt by the hot rays of the sun and die. There is no surer way of killing by kindness. A really thorough soaking, less often, does far more good and if this, or a heavy shower of rain, is followed up by a mulch around the plants, moisture will be conserved and a great deal of labour saved.

I greatly deplore the use of the hose directly on to beds and borders. The chill of tap water from the mains is highly detrimental to plants after a hot day, and the force of the water washes away the fine particles of sandy soil from around the plants, leaving them exposed. But there are now splendid devices for watering gardens. There are plastic spray-lines which are invaluable for laying along rows of vegetables, gently irrigating each row in turn. One can almost watch the crops grow as a result. There are sprinklers, whose 'artificial rain' is warmed in its passage through the air before it falls upon the plants, and is the nearest approach to rain we can give them, and they love it.

These excellent devices for keeping a garden on sand in good condition—and your garden on sand will look pretty poor without them—cost money. It is one of the drawbacks of this type of soil. It is not only in dry seasons that watering is called for; even in wet ones there occur periods when it needs to be done. We have had a succession of dry springs in recent years, and those who have watered their growing plants early in the spring know the difference it makes in size of growth and beauty of flower. The old-fashioned idea that it is detrimental to water crops early in the year dies hard, but the check which occurs, particularly in the vegetable plot, by a sudden lack of moisture, can only be harmful to young, growing plants.

Lawns. Lawns on sandy soil where the rainfall is light, are one of the worst headaches. The turf quickly becomes brown, spoiling the appearance of the garden. Watering and feeding are essential to keep grass in good condition. Lawns on acid sand are not easy, but are nothing like as difficult as on alkaline sand. Grass is difficult to establish, especially in the dry eastern counties, or on lime-soiled sand by the sea, where the most desirable grass species may suffer from drought and be crowded out by weeds. The finer grasses, such as the fescues, grow best in acid conditions so that sandy soils containing lime should be improved before sowing takes place, by working into the surface a layer of sedge peat an inch thick. The acidity may also be increased by topdressing the lawn annually with a mixture of peat and sulphate of ammonia. Special difficulties should be referred to highly expert firms of seedsmen or to the county horticultural officer.

Planting trees and shrubs. Those with experience of planting trees and shrubs on excessively light soils such as sand appreciate that this is an undertaking fraught with difficulty. It is never easy to firm sand, particularly when it is very fine, unless the sand is made very wet. Nowhere is firm staking so essential or so hard to achieve, and I have seen many instances where it was uncertain whether the tree was supporting the stake or the reverse. An excellent practice on this type of soil is to 'float' the plant into its new quarters. Quite large specimens may be successfully transplanted in this way. I have seen this carried out on very light soil in Scotland, where huge shrubs, moved only a month previously, appeared to have occupied their new positions for years. This useful, though highly expensive, operation is the quick answer for furnishing new gardens and is on the increase. This moving of mature trees, however, is a job for the specialist.

The same principle may be applied to ordinary plantings of subjects on sand. A large hole is dug out, according to the size of the subject to be planted, and filled to the brim with water. A hose is necessary where a tree or large shrub is going in. Place the shrub at the bottom of the hole with its roots spread out and shovel in soil of better quality than sand; then, as the water drains away add more water and then more soil, until only a slight hollow or depression remains. A mulch or organic material spread around the plant completes the operation. The subject which has been planted is now as firm as a rock and no amount of pulling will disturb it.

Even the transplanting of a shrub from one part of the same garden to another presents difficulties on very sandy soils. Advice to gardeners to 'lift with a ball of soil' is an impossibility on sand, the sand just drops off the spade and the uprooted plant is left, naked as a new-born babe, and has to start all over again to re-establish itself. The most one can do is to give the plant a copious watering in its old quarters the day before lifting, in the hope that some of the sandy soil may stick.

These, you may say, are small matters easily overcome, and so they may be, but the twin enemies, drought and infertility of the soil, are not so easily fought, persisting in sandy gardens unless one is continually on the watch, watering and feeding the voracious sand which is never satisfied. Even with such precautions as I have suggested, there remain many moisture-loving plants which do not succeed as on heavier, more fertile soils.

These, then, are some of the limitations of gardening on sand.

ADVANTAGES

Having faced up to the restrictions and limitations imposed by a sandy soil, we should now take a look at the very considerable advantages.

Ease of working. A well-known gardener has described it to me as 'a basic gift'. It is easy to cultivate, light and workable, not the back-breaking job it is on heavier ground, and with less and less labour available for the 'gardening lark', as I hear it is called, this is an advantage scarcely assessable in terms of time and money. Particularly is it true for the many who can only get into their gardens at the week-ends, or for the not-so-young in retirement. Indeed I feel that those about to retire, who may be moving into a new district, should make a point of discovering the kind of soil they will have to deal with, for the digging of heavy clay gets no lighter with the years.

For the keen gardener, one of the most endearing and rewarding qualities of a light soil, such as sand, is that he can get on to his land at almost any time of the year and in any weather. Heavy rains, which put the garden on clay out of action for weeks on end, do not keep the gardener on sand off his ground for more than a day or two. Showers are scarcely a deterrent at all. Thus, there is none of the frustration caused by waiting for a break in the weather while work piles up. The load can be spread evenly over the months without the mad rush to get the autumn digging finished before winter rains set in, making the ground unworkable. Indeed, on sandy soil, digging and manuring may be carried out right up to the spring. These, without doubt, are some of the reasons why many famous plant nurseries have established themselves on light soils where work can proceed smoothly and without a hitch.

Warmth. Sand, being largely composed of loose particles, holds air and therefore warmth. Owing to its dryness it warms up several weeks earlier than heavier land, so that spring sowing and planting can be started that much sooner. Though this may not be of great importance to the amateur gardener it is a vital consideration for the commercial grower who must raise and market his crops before the market becomes flooded with a glut of produce and prices drop with a thud overnight. Many a market gardener has been put out of business simply because his produce was not sufficiently early to fetch the high prices which make all the difference between a good return and a poor one. Some of the most successful market gardens and nurseries in the world are on the light sands of south-east England and the western Netherlands; no other soils produce such early fruits and vegetables.

Good drainage. Drainage on a sandy soil is excellent unless there is a hard pan immediately beneath it, which requires breaking up. This is one of the first essentials for good cultivation, since a waterlogged soil is fatal to garden plants. Excess water must be able to drain away. Many perennial plants, in particular those with woolly or silver leaves, do not survive a winter in this country on wet, sticky land, though very many come through successfully on sharply-drained sand or gravel. It is also my belief that plants in dampish situations suffer more frost damage than those on warmer, drier soils; a tender plant heavily charged with water is more likely to succumb than one growing in dry, gritty soil.

Although mature shrubs grown on poor, full-draining, light soils are considerably smaller than those grown on soils providing richer fare, the fact that they make less growth is conducive to an abundance of flower. In humid regions where rainfall is heavy, the leaching action of heavy rains gradually washes away the surface nutrients, in particular calcium and magnesium, so that the sandy soil is invariably acid. To the grower of ornamental shrubs these acid soils offer very definite advantages, since the majority of shrubs prefer an acid medium; and because of the lack of plant nutrients in acid sand, growth tends to be hard rather than sappy. This type of soil drains quickly and, most important of all, its sandy nature encourages the development of a fibrous root system which greatly reduces the risks of transplanting. Soil of this acid, sandy type tends to produce well-rooted stock, and this is yet another reason why the commercial grower of shrubs has established himself on this type of soil, while it will not have escaped the notice of the market grower that, with good management, two or even three market crops can be grown during the year on warm, sandy soil.

The way in which air permeates sandy soils allows plants of deep-rooting habit to go down deep into the ground, e.g. sugar beet on the Woburn Sands has been found to have penetrated as much as 5 feet. Very many tap-rooted plants are happiest on sandy soils, getting their moisture from far down in the subsoil.

Overwintering on sandy soil. In mild districts, where the soil is well-drained, tender plants such as dahlias, gladioli, and trigridias need not be lifted and stored in any normal winter, and losses are not so heavy as when these are lifted and stored. What a tremendous saving of labour this is for the amateur gardener who is hard put to it to find suitable winter accommodation for storing his plants. Only during the notoriously hard winter of 1962-63 did we suffer losses in our Cornish garden when plants such as these were left in the ground.

Tools. If you have cleaned tools after a hard day's work on heavy clay you can appreciate the difference between this and a sandy soil, when a wipe with an old piece of Burberry—there is nothing better—leaves the tools bright and clean ready for another day. Gardening boots and shoes last longer and do not carry into the house those wretched clods of earth so unpopular with the housewife—if housewives had their say it would be sand every time.

Uses of Sand. Pure sand has many uses. It is splendid for striking cuttings. Most of the shrubs in our garden have been grown from cuttings. Many people are unaware that the sand taken from the sea-shore above high-water mark is excellent for those plants which do not object to lime in the soil, and for many years we have used no other. Acid sand should be used for the plants intolerant of lime. Newcomers to gardening are always astonished at the ease and speed with which small pieces of shrubs, taken with a heel, will root in warm sand. It is never too early to start a 'nursery'. It should be sheltered from cold winds and from the hot sun, and it must be kept well watered during dry spells.

In winter, sand may be usefully employed for storing root vegetables such as carrots, parsnips and beetroot. Pots of bulbs may be buried under sand in the garden until roots have formed, when the pots can safely be brought into the house or greenhouse.

To sum up: you must water and you must feed a thirsty, hungry sand, but with attention to these essentials you will get results more quickly, and certainly with less effort than on a heavy soil. And if I should seem to draw comparisons between the two extremes of garden soil, sand or clay, it is only to show how fortunate I consider the gardener who finds himself on a warm, light, sandy soil.

CHAPTER THREE

Improving the Garden on Sand

'Is your garden on a starvation diet?' This was the heading of an article I saw recently in an American paper. It struck me very forcibly that this is exactly what many gardens are suffering from. We gardeners fuss and toil over plants which are generally spindly and unhealthy, and though we would be aghast, indeed infuriated, were anyone to suggest we were seriously neglecting the nutritional needs of our children, many of us grossly underestimate the requirements of our garden plants. We are content with poor, indifferent specimens which are a travesty of what they could be, simply because we disregard their basic needs. Plants depend on us for their chance in life and we must not starve them.

Is your sand acid or alkaline? There is still confusion in the minds of many gardeners between an acid soil and a sour one. Indeed, I meet with genuine affront on the part of some people when I ask if their soil is acid. 'No, indeed,' they answer 'it is very good soil, very fertile.' And this may well be so, since acid soils are some of the very best for the cultivation of ornamental plants in the British Isles, while a sour soil is one that is poorly drained and aerated, often waterlogged, so that the living organisms in the soil cannot perform their useful work. There is nothing wrong with acidity unless it becomes excessive, and even then rhododendrons and many ericaceous plants have adapted themselves to thrive in such soils.

On the other hand, there are soils in this country which are highly alkaline, and though they may be excellent for farm crops, for vegetable growing and for fruit, they are far from satisfactory for the cultivation of ornamental shrubs, many of which will not succeed under such conditions.

Soil tests. For these reasons it is essential that the soil should be tested, either doing this oneself with one of the soil testing kits available from garden stores or seeking professional advice from a firm undertaking soil analysis or the county horticultural officer.

The soil test will determine whether the soil is acid or alkaline.

This is a highly important factor, since many beautiful flowering shrubs which flourish on acid sand cannot be grown at all on one with a high lime content, for though a high degree of alkalinity is harmful to most peat-loving plants of the ericaceous type, many do not mind sand as such, but dislike the lime which accompanies some sands. There is no doubt that soil-testing takes some of the guesswork out of gardening, and there are enough unknown and uncontrollable factors so far as gardening is concerned without adding to them.

Sandy deposits away from the coast are usually strongly acid and contain a high proportion of quartz and generally a substantial amount of feldspar, but coastal sands which contain crushed sea-shells—composed of calcium carbonate—are excessively alkaline. Though acid sand is widely spread over vastly greater areas than alkaline sands in this country, it would not be safe to assume that the sand in your garden is not limy, since both types occur in many places not five miles apart. Sand with a high lime content is found most often in pockets by the sea, and it is on this most difficult calcareous sand that I garden, on the north coast of Cornwall. Not only has one to contend with the extreme poverty of the soil but the lime locks up many valuable plant foods necessary to plant growth.

ACID SANDY SOIL

The most fertile soils are slightly acid, though extremely acid soil will grow little but sedge and moss. On acid sand the gardener is fortunate, since, with certain reservations, many things will grow in it. Two of these reservations are the lack of humus and an excess of acidity.

Lack of humus. Though we do not need to go into the technicalities of the scale now freely employed for stating the acidity or alkalinity of the soil, it is sufficient to say that a pH of 7 is the neutral zone between the two extremes; readings above pH 7 indicate increasing degrees of alkalinity (pH 7·5, pH 8 and so on) while readings below pH 7 (pH 6·5, pH 6, and so on down the scale) indicate increasing degrees of acidity. The pH of sandy heathland soil can vary from about 6·5 down to as little as 4. It is my belief that a pH value in the vicinity of 5·6 is the ideal for most ornamental shrubs and trees as well as a number of other plants such as roses. Anything less than 4·5 is excessively acid and will grow little except native moorland plants like cotton grass.

Assuming, then, that the pH value of the sand is within the range of 4·5 to 6·5, preferably half-way between, acid sand will suit most plants, provided—and one cannot stress this point too strongly—that the deficiency of humus in the soil is made good by the addition of suitable materials. I stress this because I am convinced that one's efforts would lead to poor results without constant enriching of the sandy soil and it would be unfair to suggest that first-class plants can be grown on sand which is left to its own barrenness. Where the surface soil is wretchedly thin and the gardener is forced to create a soil in which his plants may grow, nothing is better than that the beds should be confined to lupins for a year or two, when all the top growth should be dug in annually. These may take the place of the annual Bitter Blue Lupin, *Lupinus angustifolius*, which Frederick the Great introduced into Germany to make good potato soil from barren farms in East Prussia, and which some British farms use today as green manure on poor, sandy or pebbly ground.

Excess of acidity. The other reservation to which I have referred is an excess of acidity. This varies, not only from one district to another, but even within the same garden and from one year to another. Any soil appreciably lacking in lime will give several adverse reactions; as the soil becomes increasingly acid, so will the incidence of pests and diseases become more prevalent. Weeds such as sheep's sorrel, spurrey and annual chrysanthemum are plants which give an indication of the lack of lime. Brassicas are attacked by club root, and some root crops, notably swedes and turnips, will be patchy in growth and may fail altogether.

Acid sandy soils deficient in lime will benefit from a dressing of $\frac{1}{2}$lb. per square yard; dressings of superphosphate will help still further to increase the fertility if used at the rate of 4oz. per square yard, and bonemeal may be applied at the same rate. On excessively acid sand, where a large quantity is needed, it should be given in several small doses as the loss by leaching increases with the size of the dose.

ALKALINE SAND

Excessively limy sand is another story. As I have already said, it occurs most often along the coast, and these most difficult seashore gardens not only have to contend with salt-laden winds off the sea, but with great poverty of soil and an excess of lime that makes it impossible to grow the superb range of peat-loving plants which form the backbone and are the glory of many famous English gardens on acid sand. Almost the only real asset of these sandy gardens by the sea, which are heavily charged with lime, is the lack of prolonged or severe frost, so that many shrubs and half-hardy perennial plants may be attempted which would not survive in inland districts on acid sand.

Lime locks up valuable plant foods and deprives plants of such nutrients as magnesium and iron, causing chlorosis or yellowing of the leaves and stunted growth. In this connection it is often useful when growing shrubs on this type of calcareous sand to underplant with some green carpeting ground cover that allows the shrubs to take up iron more fully. Gardeners on this kind of soil cannot be thankful enough for Sequestrene, based on the Geigy iron chelate Sequestrene 138 Fe, formerly Chel DP Iron, with magnesium and manganese added. This has greatly improved our hydrangeas, which showed marked yellowing of their leaves, and many other flowering shrubs which do not care for an excess of lime, including roses which previously suffered badly from chlorosis and black spot. It is a sad loss that so many beautiful plants will not grow on sand containing lime, but there are very many lovely flowering and berrying shrubs which do well, and a garden on alkaline sandy soil should make full use of them. They will be clearly defined in Chapter 4.

To attempt to bring alkaline sand back to a desirable, slightly acid region (pH 6·4), the best method is to add acid peat moss in large quantities. This can be bought with a known pH, which will do the work more efficiently, but it is never easy to make alkaline sand acid; it is far easier to counter acidity with lime.

IMPROVEMENT OF FERTILITY

Although sand is infertile, there is a great deal that can be done to bring it to a high degree of fertility—by ensuring sufficient bulk and 'body', by the addition of humus, by preventing the sand particles from drifting away with every wind that blows, and by retaining moisture and soluble salts in the sharply drained sand. Primitive peoples had none of the problems of cultivation of modern man, or rather they solved them by simply moving from one clearing to another as each plot of ground became exhausted and the crops poorer. We, however, must regularly nourish poor soil to maintain its fertility, and since the organic or humus content of sand is negligible, it must be built up and thereafter maintained at a high level to produce good crops, fine shrubs and garden flowers. First applications of organic material should be as generous as you can make them, to raise the fertility as quickly as possible; in subsequent years they may be slightly reduced but they must be continuous as the hungry sand is never satisfied. It devours whatever is put in and asks for more. Beneficial materials consist of animal manures, spent hops, leafmould, garden compost, wool shoddy and seaweed.

Farmyard manure on light sands. It is unfortunate that farmyard manure has become almost unprocurable in many districts, for it is excellent for light, hungry land, and given an unlimited supply of it, plus a liberal supply of phosphates, the gardener would have no worries about fertilisers. Manure tends to bind the sand and prevents it from falling apart into a powdery tilth. It gives necessary bulk and its first-rate capacity for holding water in soils susceptible to drought is of great value. It is also as a source of humus that it has useful results. Animal manures are valuable sources of nitrogen, and organisms increase their useful activities in manured ground. Examine a heap of dung and you will find it full of fat, red worms. Partially decayed rather than well-rotted manure is preferable on light sands, since very coarse, strawy material takes longer to decompose and is not so quickly used up. We should discard the well-worn advice to bury deeply; the place for manure on sandy or gravelly soil is in the top spit, for it is here that plants need to find the rich humus. There is no point in burying organic matter where it will lose plant nutrients to the subsoil, out of reach of the plants' roots. Although the physical effect of a single dressing of manure is slight, regular dressings are cumulative. Do not expect too much too quickly—there will be a gradual improvement over the years, through repeated dressings.

Spent hops. Where animal manure cannot be obtained, it is possible that spent hops may. Although it is one of the least known, it is one of the most valuable aids to growth on a barren sand, either incorporated into the soil or applied as a surface mulch. Rhododendrons and azeleas simply love the stuff. A gardener on Surrey sand told me he used to get 5 tons of it at a time from breweries in London, digging it into his garden or using it as a mulch, so that his lifeless, yellowish sand was completely rejuvenated and he was able to grow many beautiful shrubs and first-class perennial plants. A nurseryman on the coastal strip between Liverpool and Southport uses it in its raw state, applying it in a 2-inch layer to the surface of his nursery beds, where it acts as a mulch and helps to conserve moisture. We cannot all obtain spent hops in such quantity or afford it if we could, but we should bear in mind that it is one of the most valuable of all humus-forming materials for poverty-stricken sand, whether it is acid or alkaline.

Marling sandy soil. A century ago 'marling' or 'claying' was the practice by which farmers rendered great stretches of light sandy soils fit to carry crops. There was an old saying: 'Sand to clay will never pay, but clay to sand will make good land.' Labour was cheap in those days and time meant little, and marling was considered an excellent way of improving this type of poor soil by adding clay, or a mixture of clay and calcium carbonate where the soil was acid. Much of the light land of the eastern counties, both fen and sand, was reclaimed in this way in the latter part of the eighteenth century and through much of the nineteenth. Horse-carts brought clay from the marl pits near by—it was surprising how often clay was found in the vicinity of the sand—and they dumped it on the fields to be weathered down, so that it could be spread or ploughed-in in the spring. Today marling is no longer practised, but instead we in Cornwall see lorries carrying calcareous sand from the beaches to inland farms on heavy acid soil. Since clay is an almost inexhaustible source of plant food, those who can obtain it should spread it over the top few inches of their sandy soil and this, with the addition of some manuring, will create a medium in which most plants will grow happily.

Compost. All plants by nature are designed to live on their own rotted-down remains, so that what their leaves draw from the air and their roots from the soil one year, feeds them for the next, after it has been processed by bacteria in the soil into compost. This is Nature's way, but on poor, barren sand it is not enough and poor soil types can only be transformed into a high degree of fertility by the incorporation of compost, i.e. straw, leaves, discarded plant material such as dahlia tops and old bedding plants, spent annuals, the haulms of vegetables such as runner beans and peas and potato haulms (haulms affected by blight are not infectious once they are dead, though the tubers are, and should not be added), much of the refuse which finds its way into the rubbish-bin like vegetable remains, apple peelings, tea leaves, and even paper, which is best when damped. The compost heap is one of the most valuable assets of a sandy garden and nothing should be carried away by the dust-cart which could be added to the compost heap.

No compost heap should measure more than 5 feet wide by 4 feet high. Gardeners on small, sandy patches will be lucky if they can make as much. Begin if you can with a layer of straw. Wheat or oat straw is best. It is as cheap as anything you can buy and it is not used nearly enough. Moisten the straw as this helps decay, and then cover it with a layer of half-rotted vegetable matter. Weeds usually contain fair amounts of the plant foods needed in the soil and a special effort should be made to collect the stems and foliage of such things as nettles, valerian, chick-weed, mullein and docks. Pea haulms should be added while still green and not be allowed to wither before adding to the heap.

One of the advertised activators or sulphate of ammonia, sprinkled between each layer, hurries up the process of decay and a thin layer of soil conserves moisture. The experts tell me that sandy, gravelly soils need more than humus, they need tough and fibrous particles of rotting organic matter and advise that a good proportion of relatively hard, woody material such as thin prunings, plant stalks and bark fibre must all be included. Perhaps some of us have been too ready to take these to the bonfire and must think again.

It is astonishing how even those tiresome hard stumps of broccoli can be reduced to good vegetable compost if one keeps poultry. Put stumps in layers a foot deep and cover with strawy poultry manure to a depth of 3 inches, then add a second layer of stumps and so on. When the heap is finished, cover with straw about 1 foot deep and put some soil on top again. At the end of 12 months the great bulk of the stumps will have rotted down. Any stumps on the outside of the heap which have not completely rotted can be used to start a fresh heap.

November is the month for gathering leaves. Oak leaves are best, birch or beech not so good, beech being very slow to decay. If you can persuade the local council employees, when they sweep up the autumn leaves, to dump the leaves for you, this is ideal. A well-known gardener who gardens on sand tells me he once obtained 40 to 50 loads of leaves each autumn from this source; these were kept in a huge wire frame which also contained a batch of chickens which scratched about in the leaves, making them eminently suitable for the garden within the space of 12 months. Thus he was able to make room for the next delivery of leaves without having to keep leaves from the previous year.

Within a fairly short time he had completely altered the texture of the top soil and found that his starved garden began to take on new life and that plants and shrubs of every kind responded nobly to this treatment. This sort of massive alteration of the character of the soil is a sort of pipe-dream for most of us, as is the introduction of hundreds of tons of fresh soil which a correspondent on excessively sandy soil tells me he has recently had carried out, but even the smallest garden on sand must have its quota of leaf-mould.

Fallen leaves in the garden make a wretched mess but are, in a sense, manna from heaven. If one is fortunate enough to have a good supply one can stack them separately. Emulate woodland conditions as far as possible, making a shallow rather than a high stack, moistening each layer and adding a dusting of sulphate of ammonia to hurry things along. To prevent the leaves from blowing about, spread some old netting over the heap and peg it down. It will take some time to mature but even semi-decayed leaves are useful in a garden on sand. If leaves are in short supply add them to the compost heap, but do not, I beg of you, waste them on the bonfire.

Bracken. Bracken is excellent, particularly for calcareous sand. It is a calcifuge plant, a real lime-hater, and a sure indication of an acid soil and the height to which it grows is a guide or otherwise to the soil it grows on; on the poorest sandy heathland it rarely grows higher than a foot, though in hollows and places where humus or leafmould has accumulated over the years it will grow as tall as 8 feet. There is no question as to the value of compost made from green bracken which has been cut in June or July, stacked, and allowed to decay naturally. Such bracken is rich in nitrogen, and particularly so in potash. When it is rotted down in about three months, it will be as valuable as good farmyard manure, and is often more readily obtainable. Gardeners on sea sand should also gather it and use it as a protective mulch spread over their lime-soiled, sandy beds of shrubs and herbaceous plants during the winter. Such a surface dressing over a layer of leaves does wonders for seaside gardens with a high lime content.

Seaweed. Seaweed has all sorts of beneficial effects on the soil, and is free for the taking for those with seashore gardens. It is particularly useful in the vegetable garden and when one realises that many of our vegetables have their origin on the coast, this is not surprising. Brassicas, turnips, carrots, beetroot, and, of course, potatoes, all respond to applications of seaweed. Where will you find better crops of broccoli and potatoes than are found in Cornwall, where the ground has been manured with seaweed and the sea sand with a high calcareous content enriched by it?

It is an important supply of humus, it contains a goodly proportion of magnesium sulphate which is valuable for potatoes and for tomatoes, and it increases the power of the soil to retain moisture. As well as incorporating it into the soil, it makes a useful and beneficial mulch during periods of drought. And although it contains 2 to 3 per cent. of lime its reaction is acid, rather like that of farmyard manure, so that seashore gardeners should not fear that its use is making their alkaline soil even more alkaline. The bladder seaweeds and drift-weeds with long fronds are best.

Mushroom compost. Mushroom compost is another source of humus. The compost is inert on arrival and may be mixed with a dusting of dried blood; the compost heats up and is ready in a month. Pinks and the whole dianthus family thrive on sandy or gravelly soils into which this mushroom compost has been forked, or it may be spread around their roots during a hot summer. Mushroom compost has a good deal of lime in it, which suits pinks but would be fatal for all lime-haters.

Sedge peat. Sedge peat is especially beneficial for quick-draining light sands. It holds it together and, most important of all, it acts as a sponge and conserves moisture in the sand. No plant is put out in our sandy garden without the addition of a handful of moistened peat with a dressing of bonemeal. Plants will endure prolonged drought when given this treatment. Peat will absorb ten times its weight of water, so it is essential to water it well. Dry peat around a plant dries it and is useless.

Sewage. There is a definite prejudice against the use of sewage in this country, though it is one of the most effective additions to pure sand, and in a prepared state resembles nothing more than dry, odourless, crumbling humus. It is worth recalling that in countries with a dense population, such as China, where necessity presses hard, no fastidiousness is allowed to prevent the use of human excreta.

Green Manuring. In China, too, green manuring is regularly practised. Green manuring, or the sowing of a quick-growing crop for digging direct into the ground, is a ready means of supplying humus, when any piece of ground falls vacant and can be spared sufficiently long to allow the practice to be put into effect. Crops suitable for the purpose are turnip, rape, mustard, lupins, vetch and rye. The last-mentioned, in addition to providing humus, benefits the soil by reason of its free, fibrous root action; lupins and vetches, being leguminous crops, enrich the soil with nitrogen. Mustard is suitable for late sowing in summer, as it must be dug into the ground before it is cut down by frost.

The systematic sowing of such crops in small gardens would be extremely difficult, if not impossible, where one crop quickly follows on another, though I know of one highly successful rose-grower who never plants his roses on his poor, sandy soil without giving intended rose beds this specific treatment. The roses, once planted, never look back nor do they suffer from the diseases to which roses are prone on this type of soil.

Inorganic Foods. Gardeners must ensure their soil contains the plant foods essential for perfect health, and continuously renew these foods which plants are taking out of the soil all the time. Much of this is done by natural means which we call organic, but the food constituents into which they are ultimately reduced could equally well be provided by artificial foods. These, however, feed the plant only, whereas the organic nutrients feed the soil, providing bulk, aeration and moisture-retention, none of which a sandy soil can do without.

Three main inorganic foods are sulphate of ammonia, which provides nitrogen; superphosphate of lime, a quick-acting phosphatic fertiliser; and sulphate of potash, which provides potash. Many sandy soils are sadly deficient in potash and since one should not rely too much on the residual effect of sulphate of potash, frequent applications are necessary. Though there has been a tendency towards increased reliance on artificials in modern gardening, due to the difficulty of obtaining organic manures in urban areas, not even the most enthusiastic advocate of them recommends their exclusive use. Like stimulants, they keep things going only at the cost of increased dosage, and once they are discontinued the soil is left the poorer.

Mulching. Few gardeners make full use of mulching or surface-dressing, though it is a highly important factor, particularly with such surface-rooting plants as rhododendrons and azaleas, which suffer badly in hot, dry spells. Though it is noticeable that these grow well and regenerate naturally on heathland soils, more especially in sheltered pockets of semi-woodland areas where the deposit of dead leaves provides a natural source of humus, on open heathland which has not supported earlier generations of trees, annual mulching is the key to success on dry, hungry sand. Even some roses are grown superbly on acid, sandy soil with no more feeding than a regular annual mulch of decaying pine needles. They are never limed and do not suffer from chlorosis and very little from black spot.

Mulches may consist of spent hops, garden compost, straw, peat, semi-decayed leaves, etc., and the weed-suppression alone is worth the cost of the mulch in time and money. Mulching is normally done in the spring, though I think there is no close season for it; it can be applied with advantage in certain circumstances in the summer, but it should be put on when the ground is moist. It is a waste of time and effort to mulch dust-dry land. Since worms are most active in autumn and again in spring these are good times for them to take the dressings down into the topsoil.

6 *Above:* The long-lasting red berries of *Cotoneaster lacteus* show up well against its grey-green leaves

7 The Sea Buckthorn, *Hippophaë rhamnoides*, will grow and spread on sandy wastelands where little else thrives

8 *Olearia macrodonta* in the author's garden. It is a good hedging plant for mild situations

9 *Left: Romneya trichocalyx*, one of the parents of *R. hybrida*, flowers in late summer

10 *Below: Symphoricarpos* White Hedge is a good, compact variety with plenty of 'moth-ball' berries

I have become a complete fan of mulching, having watched the rewarding results on our sandy garden over a period of years. Small gardens are unlikely to be able to provide sufficient compost for their needs—much of it will inevitably be absorbed by the vegetable garden —but peat is easily obtainable, clean and easy to handle, and beside keeping the ground moist in summer, is an excellent labour-saver, inhibiting weeds and giving the hoe a rest. If spent hops are obtainable from a local brewery, 1 to 2 oz. of sulphate of ammonia, plus a similar quantity of superphosphate, added to each square yard of a 4-inch layer of spent hops, will rapidly enrich the topsoil.

Many gardeners have still to be converted to the technique of mulching—they continue to collect autumn leaves from the beds and borders in the interests of tidiness, even carrying them to the bonfire, burning all the precious material and robbing the plants of their natural food, and exposing the earth to the ravages of winter. No leaves, however, should be allowed to remain on grass lawns; and turf, which will quickly turn brown and show symptoms of drought within a week in dry weather, should be mulched as a preventive measure, with a mixture of peat and sifted soil, in late autumn or very early in the New Year.

A layer of organic material, not more than an inch or two thick, spread along the rows of vegetables and around the shrubs and border plants, will do all and more than we once associated with hoeing. It will stimulate the helpful living organisms in the soil, improving the soil's condition, and it should be left on the surface for the worms to take down.

Garden compost, if badly composted, will produce a crop of weeds; birds, too, are fond of pulling it to pieces in search of food, littering adjacent paths and grass verges. The tidy gardener may prefer a mulch of peat moss—it keeps down weeds and, if mixed with a fertiliser, is effective in enriching the soil as well as retaining moisture. To save mixing, scatter the fertiliser directly on to the soil before applying the peat mulch. Mulching should be continuous; every year it should be renewed until the light sand gains in colour and fertility, as a supply of grateful nourishment is gradually washed down to the plants' roots. Though the surface-rooting rhododendrons and azaleas benefit most, a mulch is excellent for most ornamental shrubs and a valuable aid in the war against black spot on roses. It should not be squandered on those plants that in Nature grow on hot, dry soils with enjoyment. These include the brooms, cistus, lavender and rosemary which come from hotter countries than ours and will flourish without any such attention.

Hoeing. Do not overdo the hoe on sandy soils. We are often told 'Hoe to keep down weeds, hoe to conserve moisture, hoe to let in air and to improve the condition of the soil.' But in recent years a large number of trials have disclosed the surprising fact that the value of the hoe is grossly overrated. Its value in the constant war against weeds is not in doubt, particularly during a hot, dry spell when the sun kills the small seedlings left on the surface, but the constant use of the hoe in the expectation that it helps to conserve moisture is now regarded with grave suspicion. Certainly it is to be deplored in the garden on sand, where it results in a greater loss of moisture than if the surface were left undisturbed.

It is also doubtful whether breaking up the surface soil on sand contributes towards improving its condition or produces better plants. Deep digging is one thing, scratching over the soil with the hoe is another. Where there are weeds the hoe must be employed regularly, but where there are few weeds there is no virtue in hoeing very sandy soil; it is better left undisturbed. Hoeing has been proved overrated as the mulch has been underestimated. Mulch more and hoe less.

Ground cover. Plants of all sorts abhor bare earth around their roots, so that for those who feel frequent surface dressings involve labour, I suggest they make use of the innumerable, spreading, ground-cover plants. Light soils exposed to the full heat of the sun rapidly lose moisture by evaporation, and by covering every inch of soil with growing plants Nature is helped and the interest and beauty of the garden is increased. Plants are not lacking which may usefully occupy these positions. The periwinkles (*Vinca*) are first-rate carpeters for light shade, even invaders such as the acaenas are excellent for clothing dry sandy banks beneath shrubs with delightful mats of coloured leaves; *Ajuga reptans* does splendidly on our alkaline soil, spreading a wide carpet of bronze leaves and short flowers of blue-purple; *Geranium sanguineum lancastriense* is a grand spreader, and the wild thyme, *Thymus serpyllum coccineus*, glows red for many summer weeks, seeding itself freely on sandy soil. *Campanula poscharskyana* may be a troublesome plant in the flower border but is immensely decorative as a ground-cover plant for poor soils in full sun. *Lamium galeobdolon variegatum* is a rampant spreader with brilliantly variegated leaves for sandy soils. All these not only carpet the ground so that weeds cannot take hold, they encourage the growth of shrubs which detest drought.

Deep cultivation. At the outset deep cultivation is essential. There is no comparison between the drought-resisting qualities of a deeply worked sand and one which has been merely scratched over. While it is impossible to grow first-class plants on sand lying over a hard, unbroken surface, the same plants astonish one when grown on a heap of loose sand. In many places sand will be found lying a bare few inches above a bed of rocky substance, and this hard pan must be broken up before a good garden can be made. It is an operation best carried out during wet weather as sand is most difficult to deal with when dry.

In cultivating calcareous sandy soil great care should be taken not to bring up the subsoil to the surface, since the tendency is for the top soil to lose some of its lime to the subsoil, and this should on no account be brought up again. Break up the subsoil by all means, to encourage the action of the worms and to enable the roots of plants to penetrate into the soil, but leave it where it is.

It is not fully realised that sand lies over a wide range of totally differing substances—clay, chalk, or even, in a few instances, over a store of water. This largely determines the plants which will succeed in any particular area. I have known moisture-loving plants such as Japanese iris, astilbes and primulas to grow well on sand where there was such an underground reservoir of water. For this reason, if for no other, it is worth finding out about the subsoil in one's garden.

Planting in difficult conditions. I have often heard members of a well-known Cornish firm, which is very experienced in establishing plants in difficult sandy conditions, giving advice on this subject and I have seen their recommendations carried out with success in gardens on almost pure sand. This advice I pass on.

Planting positions should be prepared before dry weather sets in in the spring. A suitable hole should be dug out, in which should be placed some moist peat laced with a slow-acting fertiliser such as bonemeal. The soil should then be replaced till such time as planting is to take place. If by this time, a few weeks hence, the soil has dried out a little, it must be remoistened. Plant then in the normal way, but firm with the whole weight of the body, so that the sandy soil lies in a depression immediately around the plant. Soak this depression with water and let the water soak away, then rake fresh, dry soil around the plant. Cover this soil—and this is highly important—with a mulch of chopped bracken (particularly good on sea sand), rotted manure or garden compost. I recall a newly-planted garden on sand which was given this treatment before the unprecedented hot summer of 1959, and I was astonished to see how the plants immediately started into growth and prospered.

Improving lawns on sand. A lawn which is starved of nutrients offers little resistance to weeds and in hot summer drought, turf on light soils is quick to brown. Fine grasses flourish best in slightly acid conditions, and the lawn on alkaline sandy soil is one of the major problems of this type of garden.

To reduce the alkalinity and to step up the level of nitrogen which it needs, apply sulphate of ammonia on rainy days at the rate of $\frac{1}{4}$ to $\frac{1}{2}$ oz. per square yard at six-week intervals up to the end of summer.

Some gardeners on excessively light soils are turning to chamomile in place of grass, since it remains green and in good condition when grass would brown. A Cornish nursery supplies a non-flowering strain which is excellent for the purpose. Thyme lawns of the common, wild, purple thyme, *Thymus serpyllum*, remain green when grass is browned by drought and may well be considered for gardens on sea sand.

CHAPTER FOUR

Shrubs for Gardens on Sand

I BELIEVE the keynote to success in a garden is to put the right plant in the right place, i.e. to choose a plant which will reach the height of its beauty in the soil and climate provided for it. The smaller the space to be filled, the more necessary is knowledge of which such plants are, for though our ancestors with huge gardens could make mistakes and get away with it, in smaller modern gardens our errors are pin-pointed for all to see. The best recipe for making a good garden is to discover which plants grow well in it and to concentrate on them. Simplicity of grouping, particularly where shrubs are concerned, is the thing to aim at rather than a diversity of subjects, and the limitation imposed by a smaller number of kinds would save our gardens from the crowded muddle that is often the outcome of a collection of single plants. Such a garden could be maintained with the least possible labour.

Although any natural soil will grow something, for there is an old saying: 'Where weeds will grow, other things will grow', the possible range of plants may in certain cases be very limited. Where the soil is of distinctive character the wisest course is to concentrate upon those classes of plants which are most likely to succeed. The alternative would be so to alter the character of the soil that it is made capable of growing anything we wish; this is often possible, but it is a task demanding a considerable amount of knowledge and a great deal of labour and expense. Unfortunately we gardeners have an innate tendency towards lavishing our time and money on plants which require circumstances very different from those we can offer, in spite of the fact that it is by no means always the rarest plant which commands attention; it is often one which is perfectly familiar but grown to such perfection it cannot be overlooked. By and large, it is the familiar and therefore more reasonably priced plants which make the best gardens; many plants are common simply because they are good.

The natural soil of our sandy garden was so extremely poor that it obliged us to make a special study of those shrubs which do well with the least nourishment. Any shrub which prefers a rich loam cannot be expected to flourish on thin, hungry sand, and those of us who garden on light, sharply-drained land will do well to give preference to the many delightful ornamental shrubs which enjoy these conditions rather than strive to encourage others to put up with a soil they dislike.

Fortunately shrubs are better equipped than perennial plants to withstand drought once they are established, their roots go deeper into the soil in search of moisture, and though the flowering season of those requiring a great deal of water is seriously reduced on hot, dry sand, those which have a preference for dry, sun-baked conditions are not affected.

There are many shrubs which should be given preference at the outset till soil conditions have improved, when the net may be more widely spread, according to the garden's microclimate, the local rainfall and the degree of alkalinity or acidity in the soil.

Newcomers to gardens on sand have the advantage of being able to start straightaway with the notorious sand- and sun-lovers. They will be found to be rapid growers in the warm sand they love and splendid fillers-in, while other long-term ones are maturing. Though some are not long-lived, they quickly furnish a garden with flowers and foliage, and there is no better way to improve a barren, arid sand than by planting it. Repeated croppings do wonders for poor soil types.

The following are shrubs which succeed on light sand with the minimum of care: artemisia, berberis, cassinia, cistus, colutea, cotoneaster, cytisus, escallonia, genista, halimium, helianthemum, *Hippophaë rhamnoides*, lavender, medicago, phlomis, romneya, rosemary, santolina, senecio, *Spartium junceum*, tamarisk, tree lupin and ulex (gorse).

Many other shrubs will grow on poor, light, sandy soils with annual assistance from surface mulches: for example, abelia, abutilon, amelanchier, azalea, ceanothus, elaeagnus, olearia, rhododendron and spiraea.

In preparing sandy soil for shrub-planting, it is necessary to remember that, though well-established bushes thrive without much attention, they need some help until their roots have taken hold of the soil, for more plants are lost through drought on light, dry soils during their early years after arrival from the nursery than later on. Small plants will get away more quickly than big ones, since they do not make such heavy demands on an unsatisfactory soil with little nourishment and a lack of moisture; many should be transplanted direct from pots, these being well watered in, and a mulch of some rotted material after planting will help them through their first difficult summer.

Abelia × **grandiflora.** This shrub makes a valuable 4-foot bush. It has small glossy leaves and is smothered during late summer and early autumn with pinky-white, funnel-shaped flowers and reddish bracts, which continue to beautify the shrub long after the flowers are over. Cut back a few of the older branches to the base each spring. It does well on lime.

Abutilon vitifolium. A tall, erect shrub or small tree, 8 to 10 feet high, with beautiful, grey-felted, vine-shaped leaves and clusters of large lavender flowers in June and July. There is also a very lovely white form with conspicuous golden stamens. The abutilon likes a hot, rather dry situation and is usually too tender for cold gardens, though it might be tried against a wall.

Acacia retinodes. The autumn-flowering, willow-leaved mimosa bears its scented yellow flowers on long arching branches. It has a phenomenally long season of flower, in a mild climate—from July to January. It is only suitable for mild districts or by the sea, and it likes a hot, sandy soil and shelter from cold winds. When fully grown, it can be between 20 and 30 feet high.

Afghan Sage (See PEROVSKIA).

Anthyllis hermanniae. A twiggy, deciduous shrub, about 3 feet tall, with yellow, orange-tinted pea-flowers in early summer. Excellent for a sunny spot on dry soil.

Artemisia. Many of the artemisias are from arid regions and are suitable for sunny banks of sandy soil, reaching, in general, a height of 3 to 4 feet.

Artemisia abrotanum, better known as Southernwood, has a pungent, rather fusty scent, and makes a bush of soft grey-green leaves. Cut it hard in spring. Less hardy, needing a warm soil and sun, is *A. arborescens*, one of the brightest silver shrubs I know, with deeply desiccated, lustrous foliage. Cut it hard in April to encourage new growth.

Artemisia canescens is one of the white-leaved wormwoods, a creeping shrublet which spreads a wide mat of fleecy foliage, gleaming like silver beneath a winter's sun. It is quite hardy and enjoys an arid, sun-beaten slope.

Artemisia absinthium Lambrook Silver is the best of the modern artemisias; it is a feathery mound of silver-grey with a quantity of mimosa-like spires during July.

Arundinaria (Bamboo). Bamboos, contrary to general opinion, grow well on dry as well as wet soils, though they are most prolific where the rainfall is heavy. They grow like weeds on acid, sandy soils where they can become a menace, difficult to eradicate. *Arundinaria japonica* is unsurpassed for a dense, inpenetrable screen of broad foliage carried on 7- to 10-foot canes. *A. auricoma*, *A. fastuosa*, and a near-relative of the bamboos, *Phyllostachys henonis*, are all good on sandy soil.

Atriplex. *Atriplex halimus*, the Tree Purslane, is a useful shrub for exposed sandy gardens by the sea, where it grows rapidly to 5 feet or more. Though it has no beauty of flower the satiny grey leaves are delightful. *A. canescens* is more greyish-white, of about the same height, and is another good shrub for coastal sands.

Aucuba. The aucuba is coming into favour again, after over half a century in the wilderness of neglect since its popularity in Victorian shrubberies, which was largely to blame, I feel, for the prejudice against it in this country, whereas on the Continent there are nurseries almost entirely devoted to the production of thousands of plants for which there is an unceasing demand.

Aucuba japonica has green, shiny leaves while *A. j. variegata* (*A. j. maculata*) erroneously called the Spotted Laurel, has heavily variegated foliage; each has those large, waxy red berries which are so bright in winter, if both sexes are planted. It is possible their tolerance of impoverished conditions of dry soil and their ability to survive in deep shade justify their return to favour.

Bamboo (See ARUNDINARIA).

Barberry (See BERBERIS).

Berberis. Though the yellow or orange flowers of the barberry are pretty in the spring garden it is the berries, which the family yields with such profusion, in circumstances that would break the heart of many shrubs, that are their most valuable garden feature. Berberis are splendid berrying shrubs for dry banks of sandy soil and many have most beautiful autumn colouring. They seed freely on our limy, sandy soil but do not come true from seed, so that any particular variety must be propagated vegetatively. They are bad transplanters and must be planted when quite small.

Berberis × *stenophylla*, with long whiplash wands, makes a fine hedge but is altogether too overpowering for the small modern garden. The semi-evergreen *B. wilsoniae* has brilliant autumn colour and a mass of translucent coral-red berries, and is a compact shrub of 4 feet. The berries are slow to go as the birds do not care for the prickly spiny leaves. Other good kinds are: *B. verruculosa*, *B. linearifolia* Orange King, *B. aggregata* Buccaneer and the small *B. buxifolia nana*. This last and *B. gracilis* (now *Mahonia gracilis* but usually listed by nurserymen as *Berberis* × *stenophylla gracilis*), and the delightful small *B.* × *stenophylla corallina nana,* are attractive for small gardens. Berberis are all tolerant of lime and are invaluable for dry, sandy gardens.

Beschorneria yuccoides. This most curious plant from Mexico, with great arching flower-spikes rising from rosettes of sword-like leaves to a height of about 4 feet, is too tender except for mild districts or by the sea. The strange green flowers, lightened by rose-red bracts, hang rather like fuchsias down the length of the long, curving spikes in June. It is not difficult to establish on sharply-drained, sandy soil in full sun. An unusual plant for coastal gardens.

Bladder Senna (See COLUTEA).

Box (See BUXUS).

Broom (See Chapter 7).

Buddleia. Though buddleias of the *davidii* type find our limy, sandy soil lacks sufficient moisture for their needs, these are splendid shrubs for sandy gardens where the rainfall is heavier. They should include such varieties as the red-purplish Royal Red, the deep purple Black Knight (the darkest of all buddleias), the violet-blue Empire Blue, White Profusion, with massive white plumes and abundant side flower-spikes, and the lilac-pink Fascination. They grow about 9 to 12 feet high and hard pruning in early spring produces the best flowers in July and August.

Two compact buddleias, in flower a little earlier, are *B. fallowiana alba* and *B. f.* Lochinch. *B. f. alba* is intensely white in all its parts, with silvery-grey foliage and white flowers with an orange eye, while the beautiful Lochinch has the bluest flowers of any buddleia, and blue-grey leaves. They put up with drier conditions on sand than those of the *davidii* type. The graceful *B. alternifolia* may be seen on the light, hungry soil at Wisley trained as a weeping tree, its pendulous branches wreathed in soft purple flowers. *B. globosa,* with rounded orange balls in May and June, may be grown on almost pure sand, thriving in full coastal exposure. The newer *B. g.* Lemon Ball, only recently introduced by a Cornish nursery, has lemon balls of flower, well-spaced out on tall, rather stiff stems, and will be a welcome addition to gardens on sand when it becomes better known.

Bush Laburnum (See PIPTANTHUS).

Bush Clover (See LESPEDEZA).

Butcher's Broom (See RUSCUS).

Buxus sempervirens (Box). Edgings and small hedges of box were familiar sights in old gardens but are rarely used in modern gardens, since they need constant clipping, and provide homes for snails and slugs. I have seen good clipped topiary specimens in gardens on sand.

Calceolaria integrifolia. This vigorous dwarf shrub, covered all summer with clusters of yellow, pouched flowers typical of the calceolaria, loves a sunny situation on sandy soil, but is only hardy enough for mild localities. Take cuttings in late summer and keep in a sandy frame as replacements.

Californian Tree Poppy (See ROMNEYA).

Calluna (See Chapter 7).

Cape Figwort (See PHYGELIUS).

Caryopteris. Those who have seen the wild form of this small shrub on remote beaches in Greece, with its roots in pure sand, will know that these are the sort of conditions this pretty shrub enjoys and will plant it in gardens of lime-soiled sand. *Caryopteris × clandonensis* has narrow, greeny-grey leaves and dainty, violet-blue flowers in August. Ferndown is a deeper blue, and Kew Blue is a smaller shrublet of dark violet-blue. Prune them hard in early spring. They are first-class, small, 3- to 5-foot shrubs for modern gardens.

Cassia. A shrub for a sunny wall in mild localities only, *Cassia corymbosa* bears large clusters of butter-yellow flowers from July onwards and is easily raised from seed. *C. acutifolia* is a tall, erect bush which has lemon-yellow clusters of flowers late in the year, and may be grown in the open garden in mild, sheltered situations. Both reach a height of about 5 to 6 feet.

Cassinia. *Cassinia fulvida* is a heath-like shrub, 5 to 6 feet tall, for limy coastal gardens on sand, and though the tiny white flowers are insignificant the golden effect of the whole bush is delightful against shrubs with darker foliage. It is similar in appearance to *Olearia solandri* but more frost-hardy, coming through recent severe winters unscathed.

Cassinia leptophylla is also heath-like but of an altogether whiter appearance. Cassinias respond to quite severe pruning after they have flowered in summer.

Ceanothus. Though the winter of 1962-63 took heavy toll of these tender shrubs from California, they provide some of the best displays of blue to be found among shrubs in this country, particularly on soils with a high lime content where the blue hydrangeas cannot be grown. That they come from stony hillsides of the Pacific coast provides a pointer to the conditions they enjoy; they do not want a rich loam— indeed they give their most prolific show of flower where the root-run is restricted and where the ground is sharply drained. Ceanothus are sun-lovers and though many are hardy enough for warm coastal gardens, they should be given the protection of a wall in colder districts. Deciduous varieties are more frost-tolerant than the beautiful evergreens but all are susceptible to wind-rocking and slabs of stone placed over the roots will ensure that the plants do not rock in the ground.

Ceanothus thyrsiflorus is one of the toughest and hardiest of the evergreens, though I find the newer and more compact *C. t. prostratus*, a low hummock of a bush with rich blue puffs in spring, more stable in windy positions. *C. dentatus impressus* is another spring-flowerer among the evergreen ceanothus, which is hardy enough for a sunny slope or bank in the open garden, with very distinct small leaves on stiff shoots and deep blue flowers in May. From July till autumn *C.* Autumnal Blue will be laden with spikes of soft blue, on a fast-growing bush to 8 feet; it is one of the hardiest of the evergreens.

All ceanothus are resentful of root disturbance and should be planted direct from pots. Spur back the long shoots of the evergreens as soon as they have flowered, usually in July, to allow the young wood to harden before bad weather sets in; Autumnal Blue should be cut hard back in April and deciduous varieties in March.

Centaurea gymnocarpa. This drought-resistant plant makes a dwarf bush about 2 feet high of silvery ferny foliage and its purple thistles in July are entirely secondary to the beautiful leaves. It is not dependably hardy and cuttings should be taken in early September in a sandy frame, or seed may be sown in August in the greenhouse.

Chaenomeles. The ornamented quinces, previously known as *Pyrus* or *Cydonia*, or even simply as Japonica, are lovely spring-flowering shrubs for the open garden as well as against walls, and they are in demand on limy soils which preclude the planting of azaleas. Knap Hill Scarlet, 5 feet by 7 feet, the prostrate *simonii*, with blood-red flowers, and the bushy Cameo with flowers of soft peach are most beautiful.

Choisya ternata (Mexican Orange Blossom). This rounded shrub of glossy leaves has clusters of exotic-looking white flowers in May which scent the air. It does well on light sandy soil, though it repays some feeding and can reach a height of 6 feet or more. Plant with shelter from cold winds. It is reasonably hardy except in the severest winters.

Cistus (Rock Rose). No genus is better equipped to cope with the impoverished conditions of poor, dry soil than the cistus. The wild species come from Mediterranean regions and are specially abundant in Spain and Portugal. Few shrubs respond more eagerly to the sun's warmth, renewing their fugacious, saucer-flowers each day under its influence, though few shrubs suffered more from our recent notoriously hard winter than they. This aromatic family can be relied upon to flourish on sandy or gravelly soils where little else will survive. Hot, sandy banks in full sun with shelter from cold winds, provide ideal conditions and they are excellent for seaside planting. No shrub is fonder of the company of its fellows, and it pays to plant two or three together, as this is one of the greatest deterrents to frost damage that one can give them. Since only a few are hardy in our climate, cuttings should be taken in a frame in July or August.

Probably the most reliably frost-hardy is *Cistus laurifolius*, a tall 6-foot shrub with dull green leaves and a profusion of white flowers carried erect in June; it will thrive on almost pure sand if deeply dug. Almost as hardy, though more beautiful, is the Gum Cistus, *C. ladaniferus*, a tall bush of dark leaden foliage, which is intensely gummy, and very numerous white blossoms marked with a deep maroon-red blotch on each petal—a glorious sight for many summer weeks, and it never flowered better in our garden than after its baptism of ice and snow during the winter of 1962-63.

Cistus crispus, a low-growing species, is rarely without a few of its rose-magenta flowers even after the main June flush is past, and its delightfully crinkled grey leaves, so resistant to salt winds, are a decoration all on their own. It hybridises so freely that the flowers on different bushes are extremely variable in colour. *C.* × *corbariensis* is as hardy as any of the low-growing cistuses; it rarely achieves more than a couple of feet and a bush is covered with a mass of white saucer-flowers about the size of a shilling.

For seaside exposure I find *C.* × *aguilari maculatus* hard to fault; it has light green leaves, waved at the edges, and enormous, creamy-white flowers with purple markings at the base of the petals, on a bush 4 to 5 feet tall and with as wide a spread. *C. purpureus* is more tender but the large rosy-crimson flowers, with a deep maroon blotch, are so beautiful that it should have a warm corner on very poor, stony ground, where its chance of survival will be greater. Silver Pink is a cistus with no shade of puce in the shell-pink flowers, but it needs frequent nipping of the young shoots to retain a neat, rounded bush. I have a preference for the sage-green leaves and soft pink flowers of *C.* × *skanbergii* with its resemblance to a dog-rose. It makes a low bush, 3 feet by 3 feet, and has proved hardier and longer-lived in our garden.

One of the loveliest of small cistus is *C. palhinhaii*, discovered among the granite rocks of Cape St Vincent. It has extraordinarily dark sticky leaves, which are subject in some gardens to a sooty mould, and enormous, pure white, crinkle-petalled flowers with bosses of golden stamens. It is not as frost-hardy as one could wish and the presence of the sooty mould must detract from its value as a permanent small shrub.

Cistus Paladin (*palhinhaii* × *ladaniferus*), whose white flowers, marked with maroon, are the largest of any cistus I have seen, is a beautiful hybrid and deserves a prominent place in the front of the shrub border. Nip back the young shoots to maintain a 3- to 4-feet bush.

Clematis. Although I do not recommend the large-flowered hybrid clematis for sandy gardens, particularly on acid sand, as they seem to need a richer diet than a sandy one and dislike an excess of acidity, vigorous growers such as the white-flowered species *Clematis montana* and the pink form, *C. m. rubens*, quickly cover and outgrow their allotted space. These spring flowerers are splendid climbers for clothing fences or the trunks of dead trees. The winter-flowering *C. calycina*, the fern-leaved clematis, with hanging bells of cream, spotted inside with red, and ornamental foliage, does well on quite poor soil and will stand coastal exposure. The pretty *C. macropetala*, with violet-blue flowers in May and June, does well on our limy sand, and *C. tangutica* will be extravagant with yellow hanging bells in late summer and delightful shaggy, silver seed-heads in a sunny place. All clematis respond to plenty of moisture in the soil.

Clerodendron. This is one of the most exciting berrying shrubs for sandy soil where there is lime. *Clerodendron trichotomum fargesii* scents the August garden with its white flowers, but it is the unusual, turquoise-blue fruits which are outstanding. It is slow to flower in infancy but is worth waiting for. It suckers rather too freely and should be carefully sited on this account. Water copiously during the growing season and give a mulch of peat. It reaches a height of up to 10 feet.

Colletia armata (Mexican Bodkin Bush). This 8-foot bush has branches densely packed with green spines and is covered in autumn with profusions of tiny pinky-white flowers, which are fragrant like the hawthorn. To encourage free flowering give it a sunny, sheltered position on sandy, well-drained soil where it will prove reasonably hardy.

Colutea (Bladder Senna). *Colutea arborescens* and *C. × media* are excellent for furnishing dry banks of sandy soil, and though by no means among the first rank of choice shrubs, are useful for extremely poor, arid sand where little else will grow. Their copious branching and bushy habit, the pinnate leaves, and strange bladder pods which follow the yellow or bronzy-yellow flowers are not without interest. They flower late and for a long time, but take up considerable room. Prune them nearly to the old wood in February. When fully grown they may be 8 feet high.

Convolvulus cneorum. A delightful small shrublet, not more than 2 feet high, with silvery-grey leaves of pure satin and clusters of pink buds which open to pinky-white, funnel-shaped flowers at midsummer. It is only hardy on a light sandy soil and is excellent by the sea in full exposure to salt drift.

Cordyline australis. The New Zealand Cabbage Palm grows vertically to 30 feet or more in extreme exposure and gives a sub-tropical effect, with tiny scented white flowers in large branching clusters, but it is hardy only in mild localities or by the sea on light, well-drained soil. It is most often seen in public gardens or on the fronts of seaside resorts. This plant was formerly known as *Dracaena australis*.

Cornel (See CORNUS).

Cornelian Cherry (See CORNUS).

Cornus. Many of the Cornel family prefer a moist, moderately rich loam, but the Cornelian Cherry, *C. mas*, will make do with less rich fare. It is a small, spreading, deciduous tree of bushy, vigorous growth, whose masses of bright yellow flowers are borne on naked twigs in early spring.

Corokia. *Corokia cotoneaster* is a dwarf, rounded shrub of tortuous, interlacing branches, rather sparsely covered with tiny, spoon-shaped leaves and starry yellow flowers in May, followed by red berries. Its strange branching system never fails to attract attention and is greatly enhanced by an underplanting of some subject with more solid foliage, such as one of the bergenias. *C. virgata* is more erect, with small bronze-green leaves and golden stars and orange-yellow berries. These New Zealanders suffer in hard winters in this country but are reasonably hardy on light soil in the south.

Coronilla. *Coronilla glauca* should be in every garden on light sandy soil where frost is not severe. The rounded bush of pretty glaucous leaves is gay with clusters of yellow pea-flowers all winter in the mild south-west, while in colder counties it flowers from spring to mid-summer. Pinch back the tips of shoots after it has flowered to keep it bushy at about 4 or 5 feet high, and take cuttings in summer in a frame as replacements after a severe winter. Give it a dry position with shelter from cold winds. *C. valentina* is dwarfer with finer grey foliage.

Like many of the *Leguminosae*, *C. emerus*, does very well on hot, dry sand, coming through recent severe winters when other shrubs succumbed. It is an elegant shrub of vetch-like leaves and yellow flowers marked with reddish-brown. These are produced in such profusion that a bush is a pretty sight in May and again in autumn. Young plants come well from seed or cuttings. It is known as the Scorpian Senna.

Cortaderia (Pampas Grass). There are good named cultivars of the common Pampas Grass, *Cortaderia argentea*, which are vastly superior to the type. *C. a. rendatleri* and Sunningdale Silver are both good. Though I doubt if many of us have the courage to set fire to last year's foliage in the spring, this is the best way to deal with those long rigid spikes of faded silvery plumes, simpler and better than cutting them back. Plant in April. Although not shrubs, these are often classed as shrubby plants and planted as such.

Cotinus. *Cotinus coggygria*, the Smoke Plant, will be known to most gardeners under its old name, *Rhus cotinus*, and is still listed under that name in many catalogues. This species, like others of its kind, is easy to please in respect of soil, growing well in poor, dry ones. Its common name derives from the dense feathery inflorescences which are borne in summer. In autumn the leaves turn brilliant yellow, the colour being best when the plant is grown in sandy loam. *C. c. foliis purpureis* Notcutt's Variety is a great favourite for the dark maroon foliage, associating wonderfully with other shrubs, and it is one of the best shrubs with purple leaves. They prefer an acid soil.

Cotoneaster. A very varied group of shrubs, evergreen and deciduous, which ask little from the soil and succeed in very dry situations. They seed themselves freely on our very alkaline sand. There are tough and hardy ones for almost any situation, from the prostrate *Cotoneaster dammeri* and *C. microphyllus*, which is perhaps the toughest of a tough race, for it tolerates the windiest exposure and the poorest of soils, to *C. horizontalis*, the well-known fish-bone cotoneaster, so useful for concealing ugly places, and the beautiful *C. conspicuus decorus*, with a height of no more than 2 feet but a spread of 6 feet. The last-mentioned is a smother of white in June and it bears brilliant red berries in autumn which last well, since the birds seem to leave them alone. Cotoneasters with yellow berries make a pleasant change and the tall, graceful *C. salicifolius fructu-luteo* makes a contrast to the scarlets and reds. Good hedges are made from *C. simonsii*, gay with orange-red berries even when restricted to a 3-foot hedge, and the tall *C. lacteus*, which berries late and whose matt crimson fruits last a long time. It is also excellent trained against a wall. Those with lime in their sand should make use of these rewarding shrubs.

Curry Bush (See HELICHRYSUM).

Cytisus (See Chapter 7).

Daisy Bush (See OLEARIA).

Daphne. Though the daphnes have a reputation for 'miffiness' *Daphne odora aureo variegata* does splendidly on our limy sand. As its name implies, its leaves are heavily margined with yellow, while the soft reddish-purple flowers have the most delicious scent. It is one of the joys of our January garden. *D. mezereum*, capricious as it is, sometimes thrives under hot, dry and poor conditions and in fact often propagates itself freely by seed; it is an uncertain plant and may fail completely on what would appear to be an ideal soil. Both the above are slow-growing, to about 3 feet.

Deutzia. Two deutzias which may be planted with confidence on rather poor, light soil are *Deutzia pulchra*, a spreading, 6-foot bush of grey-green foliage and a profusion of panicles of pearly-coloured flowers in June, and *D. scabra candidissima* whose double, pure white flowers are carried on stiff stems which splay out from the base in a vase-shaped fan. I do not know Codsall Pink, whose double flowers are rose-purple, and whose habit is said to be similar to that of *P. s. plena* (Pride of Rochester), but I hear good accounts of it on sandy soil where other deutzias fail for lack of moisture.

Dracaena (See CORDYLINE).

Elaeagnus. These tough and wind-tolerant shrubs include many with beautiful foliage, which makes the genus extremely valuable, particularly on limy coastal sand, where evergreens are difficult. *Elaeagnus pungens aureo-variegata* is one of the most striking of all winter shrubs and quite the showiest of those with variegated leaves. Give it a sunny place to light up the beautiful leaves, with a bright gold patch in the centre of the bright green leaf. It is excellent as a cut foliage plant. *E. p. dicksoni* and *E. p. fredericii* are also good.

   The deciduous *E. argentea* is a wonderful drought-resister for a dry, sandy bank where it glistens with silver. *E. macrophylla* is a shapely bush of rounded leaves shot with a glistening metallic lustre and silvery-white beneath. *E. glabra* has its uses as a wind-stopper among pines which have lost their lower branches; it throws out long shoots which are capable of climbing up adjacent trees. *E.* × *ebbingei* is a first favourite for sandy coastal gardens exposed to salt winds, with dark important foliage with a silvery reverse, on a tall fine shrub. It can

11 The white 'immortelle' flowers of *Anaphalis triplinervis* can be dried and used for winter decoration

12 Another view of the author's garden, with *Anthemis cupaniana* in the foreground

13 *Above left:* The yellow-flowered *Anthemis tinctoria* grows to about $1\frac{1}{2}$ feet and follows *A. cupaniana* (illustrated on the previous page) in flower

14 *Above right: Echinops ritro* is a 3-foot-tall, thistle-like plant, with steely-blue flower-heads on stiff stems which need no staking

15 There are several forms of sea holly, all very much at home on sand, and *Eryngium giganteum*, shown here, is particularly beautiful

16 The individual flowers of Day Lilies are not long-lived, but a group of *Hemerocallis* hybrids provides a more lasting display

scarcely be bettered as a decorative shelter shrub by the sea. Elaeagnus are slow-growing to about 8 or 10 feet and, except for the variegated *E. pungens aureo-variegata*, are not reliably frost-hardy. They are easily propagated from cuttings.

Erica (See Chapter 7).

Erinacea pungens (The Hedgehog Broom). A spiny, dome-like shrub, no more than 12 inches high, with pale violet-blue pea-flowers in April and May. It prefers a dry, sunny place. It is sometimes listed as *Anthyllis erinacea*. It is not very common as it is difficult to grow from cuttings.

Escallonia. This genus includes some of the most wind-hardy evergreens for dry, sandy soils in the southern half of England, which are among the very best flowering shrubs for coastal planting, since they put up with driven salt-spray off the sea. For extreme exposure the gummy-leaved *Escallonia macrantha* has bright red flowers and a strong fibrous root system which allows it to withstand the most violent gales without rocking, though a better garden hedge plant is Crimson Spire, with dark foliage and bright crimson flowers. Crimson Spire makes a narrow 6-foot hedge in four years in favourable conditions. Donard Seedling is a pretty shrub of arching shoots, studded with rose-pink flowers that turn white with age, and the very hardy *E. × langleyensis* has bright rose-crimson flowers all down the slender, cascading branches. The hardy *E. × exoniensis* grows with astonishing vigour on poor sandy soil, making a sturdy bush of white flowers flushed with pink, while the lovely *E. × iveyi* is pure white against very dark glossy leaves.

The Apple Blossom series is a race of compact, bushy escallonias which suit the modern garden. While the rich red Pride of Donard is the first to flower, it is soon followed by the large, soft pink clusters of Apple Blossom, and the newest and the best of this famous series, Donard Star, whose rose-pink flowers are the largest of all the genus and which is reputed to be the hardiest of these attractive hybrids.

All escallonias are tolerant of lime; they prefer a light soil on the dry side; no manure should be added when preparing the sites or as a mulch afterwards. Recent severe winters have enhanced their reputation for hardiness, since many which appeared moribund are growing strongly again. Pruning is essential to maintain an attractive bushy habit and should be carried out in June or July immediately they have gone out of flower. Hedges which have grown too tall and wide may safely be cut back severely. They will break anew. Cuttings taken with a heel of old wood are easily rooted in late summer.

Euonymus. *Euonymus japonicus* is one of the toughest evergreens we have for shelter planting on coastal sands in the southern half of the country. It is much loved by cattle and though it is quite innocuous to them—indeed farmers tell me that the fat content of milk is greatly increased when they have eaten it—this is not its purpose. It grows slowly to about 15 feet, and owing to the very dense fibrous root system it may safely be moved when quite large. There are good variegated forms such as *E. j. albo-marginatus* and *E. j. ovatus-aureus*.

Eupatorium micranthum. The Mexican Incense Bush is a tender shrub for favoured localities or by the sea, where it quickly makes a rounded bush up to 5 feet high, smothered in August with flat, white flower-heads, reminiscent of the gypsophila. It is worth taking cuttings to overwinter in a sandy frame, since young plants flower at an early age.

Fabiana violacea. This heath-like, low-growing shrub is an interesting and beautiful evergreen from Chile, of dense, spreading habit, with sprays of greyish-green, juniper-like foliage and soft mauve, tubular flowers in June. It is unfortunately rather tender, but is hardiest on light, dry soil in a sunny, sheltered position. It is not lime-tolerant.

Fatsia japonica. This strange shrubby plant, often grown as a house plant, is entirely at home in sandy gardens by the sea. The huge glossy leaves and creamy inflorescences followed by black fruits make it valuable for decorating a corner position where a bold subject is required, and there is room for it to grow to 6 feet or more.

Figwort, Cape (See PHYGELIUS).

Firethorn (See PYRACANTHA).

Fuchsia. There are few soils in which the hardy fuchsias cannot be established, from the light acid sand of the Bagshot district to our own limy coastal sand. I would not hesitate to try them anywhere where the climate is not too harsh. They succeed admirably, with very little attention. As well as *Fuchsia magellanica riccartonii*, an excellent hedger, try ones with larger flowers such as Mrs Popple, Lena, Madame Cornelissen, Margaret and Mrs W. P. Wood, all of which have survived recent winters in the open garden, when given a winter covering of peat, leaves or ashes.

Garrya elliptica. This vigorous shrub, which may be grown against a wall or as an open bush, has matt, grey-green oval leaves and hanging catkins of silvery-grey from November until February. Only the severest winter will damage it. Plant the male form, whose catkins are longer and finer. The best time to prune is immediately it has flowered, before fresh growth begins, and it can be kept to about 6 feet or so, as desired. It prefers an acid soil.

Genista (See Chapter 7).

Gorse (See Chapter 7).

Griselinia littoralis. This wind and salt-tolerant evergreen with pale green, fleshy leaves does not care for the dryness of our limy sand, though I have known it reach tree-like dimensions on almost pure acid sand.

Gum Cistus (See CISTUS).

Halimiocistus × sahucii. This prostrate shrub, with a huge spread, has dainty, deep green foliage and wide, white saucer-shaped flowers borne in great profusion in June. It is a cross between *Halimium* and *Cistus* and is quite hardy enough for most parts of the country on a light, dry soil. Regrettably less hardy is the beautiful *H. × wintonensis*, with larger white flowers, with a zone of crimson-maroon and a yellow patch at the centre—the flowers remain open after midday. It is well worth a trial on a sunny bank in a warm garden and replacements are easy from cuttings.

Halimium. There is no yellow cistus, though the halimium closely resembles one, and enjoys similar conditions of sandy soil and a place in the sun. The best is *H. lasianthum* (*formosum*) whose large yellow flowers, with a brownish-purple blotch at the base of the petals, cover a spreading bushy plant of grey downy foliage from May to July. *H. ocymoides* is very similar though less showy.

Halimiums have a liking for lime in the soil and bloom most profusely in full sun.

Halimodendron argenteum. The Salt Tree from the dry salt-fields of Siberia is a tall, 6-foot shrub with silver-grey pinnate leaves and pea flowers of purplish-pink in summer. It is quite hardy and happy on very sandy soils or by the sea.

Hebe (Veronica). I would not hesitate to try any of the hebes on a sandy soil, though a few are more sand-tolerant than others. The amazingly tough varieties of *Hebe speciosa* in a wide range of colours endure poor conditions of soil and extreme exposure at the sea's edge in a way no other shrub will do. The hummocky *H. dieffenbachii* seems indifferent to poverty of soil and I have seen it give a grand display of its mauve-white flowers on a dry wall composed mainly of sand. The very hardy Miss E. Fittall should also be mentioned as putting up with extreme poverty of soil, and does not fail to produce the long, slender spikes of lavender-blue flowers from June to November. Though these are perhaps the most tolerant of excessively sandy conditions, many others will endure with fortitude conditions too dry and poor for other shrubs, if climatic conditions are not severe.

Hedgehog Broom (See ERINACEA).

Helianthemum (Sun Rose). Smaller than the cistus or halimiums, the bright helianthemums provide mats of varied colour in the front of a border on poor sandy soil in full sun—indeed the poorer the soil the better they flower. I like to visit a nursery where they can be seen in pots, for there is an enormous variety of colour. Double-flowered sun roses have much to recommend them. Being sterile they go on blooming all summer. I like especially the varieties Jubilee, canary-yellow, and Cerise Queen. Almost the only fault of the sun roses is that after a few years they get old and straggly and need replacement.

Helichrysum. It is sad that many of these charming, low-growing silver-leaved shrubs are so tender and intolerant of excessive wet, though they are beautiful on sandy soil which is sharply drained. They flourish on any sand, acid or alkaline, and even grow in extreme coastal exposure.

Helichrysum angustifolium is the Curry Bush, with narrow grey leaves and mustard-yellow, tiny flowers on erect stems in late summer, and the foliage and twigs have a pungent aroma reminiscent of curry. *H. plicatum* has narrow, spiky, grey leaves and golden flowers on slender stalks. *H. stoechas* is a most decorative plant of long arms clothed with white down and tiny grey leaves. It came through the bitter winter of 1962-63 in our garden on a desiccated bank of poor soil. *H. petiolatum* will certainly be killed by hard frost, but its woolly grey-white leaves are so enchanting that cuttings should be taken of this fast-growing plant and set out in May to make wide spreads by mid-summer. The flowers, carried on long stems, are off-white.

Hippophaë rhamnoides (Sea Buckthorn). Few shrubs are better equipped to bind shifting sand or so decorative for mass planting on sand dunes. Its suckering habit makes it entirely suitable for planting rough, sandy wasteland where little else will grow, and its narrow grey leaves and bright orange berries, if one male has been planted to windward of a group of females, are most pleasing. It can grow to between 15 and 25 feet though 8 to 10 feet is more usual on sand.

Hoheria angustifolia. I have been astonished at the vigour of this lime-tolerant shrub or small tree on very sandy soil. It resembles an evergreen weeping birch, with cascades of fragrant, cherry-like, white blossoms during late summer. Though undemanding as to soil it likes some moisture.

Honeysuckle (See LONICERA).

Hydrangea. Though hydrangeas are difficult on our dry, limy sand, on light acid sand in parts of Surrey they often do extremely well with very little help when planted. They prefer a *p*H of 6·5 or less and do best where the rainfall is heavy or the ground is not allowed to dry out.

Hypericum calycinum (Rose of Sharon). Never plant this very invasive shrub where its spreading tendencies may be a menace. It is most useful for covering awkward situations where little else can be induced to grow. Cut over the old growth in early spring with a sharp pair of shears and the young fresh leaves will make a pleasant setting for the large, golden, saucer-flowers at mid-summer.

Hypericum × moserianum is a grand old shrub, low-spreading with short arching branches, flowering luxuriantly all through the late summer and early autumn, and entirely carefree on a light soil. The dense bushy habit of *H. patulum* Hidcote and the profusion of its large, golden saucer-flowers makes this a most worth-while garden shrub. Do not be afraid to prune last year's growth to its base each spring. Though hypericums are not averse to lime, they do better on the moister acid sand than on the drier alkaline one.

Hyssop (See HYSSOPUS).

Hyssopus aristatus. Hyssop is a small, evergreen, hardy shrublet, about a foot high, that deserves to be better known in gardens. A row of hyssops makes an informal low hedge of rich green foliage and profuse spikes of purplish-blue from June onwards, and the flowers are nearly always alight with bees. It comes easily from seed or may be propagated from cuttings. Give it a sunny position on well-drained sandy soil and trim it over in the spring.

Jerusalem Sage (See PHLOMIS).

Juniper (See JUNIPERUS).

Juniperus (Juniper). There are good hardy junipers for gardens on sand and, since they are lovers of a limy soil, they are welcome ever-greens for seashore gardens. The cultivars of *Juniperus squamata* include some of the most distinctive. *J. s. meyeri*, the bluest of them all, is perhaps deservedly the best known. It slowly attains a height of 5 to 6 feet and has a bushy habit that makes it one of the most attractive in the garden, while the slow-growing *J. s. wilsonii*, of a light silvery-blue, is a beautiful specimen for small gardens.

They are pleasing accents among the dwarf prostrate kinds such as *J. × media pfitzeriana*, no more than 3 feet high with a wide spread, *J. m. p. aurea* whose foliage is faintly tinged with gold, or *J. chinensis variegata* whose shoots have white leaves among the green. Junipers are increased by short cuttings of the current year's shoots in a sandy cold frame in autumn.

Kerria japonica. Though the double-flowered *K. j. flore pleno*, with orange-yellow, ball-shaped flowers, is the better known, the single-flowered species, with pretty saucer-shaped, golden-yellow flowers, has a graceful, bushy habit about 4 or 5 feet high and is rarely without a flower.

The kerrias do not need a rich soil and since they come into flower in late spring a sandy soil does not lack moisture so early in the year.

Laurustinus (See VIBURNUM).

Lavandula. No sandy soil is too hot or dry for the sun-loving lavenders. They are most happy on a calcareous soil. As well as the Old English Lavender, *Lavandula spica*, good dwarf kinds are Munstead Dwarf and Twickle Purple. Lavenders are not long lived and need annual clipping after they have flowered to keep them shapely. The very distinct *L. stoechas* is more tender and should have the hottest, driest spot in the garden—the top of a dry wall enables one to see and admire the unusual deep purple, bottle-shaped spikes. It is a pleasing contrast for the brightly-hued gazanias.

Lavatera. Few shrubs grow faster than the Tree Mallow. *Lavatera olbia rosea* has bright rose flowers, like single hollyhocks, on a tall 6-foot bush of soft grey-green leaves. Like all the mallow tribe it seems indifferent to wind or poverty of soil. Shorten the long stems in the spring and it will flower from June far into the autumn. The less common *L. assurgentiflora* has grey, vine-shaped leaves and large, pale mauve flowers with conspicuous purple stamens, from May till late summer. It is not reliably frost-hardy even in the south-west and is best in the shelter of a wall.

Lavender (See LAVANDULA).

Lavender Cotton (See SANTOLINA).

Leonotis leonurus (South African Lion's Tail). Any plant which flowers in late summer or autumn is worth considering. This striking tall shrub with erect shoots which bear large whorls of downy orange flowers from August to October is rather tender and is best in a sheltered position against a wall or hedge in a favoured locality, when it will grow to about 3 or 4 feet. It needs a well-drained, sandy soil, for it will not put up with excessive wet or bad drainage.

Leptospermum. The New Zealand Tea Tree or Manuka is only suitable for mild, frost-free districts such as Cornwall or the south of Ireland, on a light sandy loam free from lime. It is a genus of shrubs with small heath-like leaves on slender arching branches. The Crimson Manuka *Leptospermum scoparium nichollsii* has tiny purple leaves and rich crimson flowers in June while the Boconnoc Form flowers twice, in June and again in December.

Leptospermum flavescens obovatum is white-flowered, with tiny bright green leaves, and Red Damask is new and beautiful with double flowers of cherry-crimson.

Lespedeza thunbergii. Often called the Bush Clover, owing to the clover-like trifoliate leaves of this leguminous plant, this is a semi-woody shrub, 4 to 6 feet tall, with rose-purple pea-flowers produced abundantly in late summer. It is hardy in a sheltered position on a light soil, and is worth trying for its show of flower late in the season. It enjoys similar situations to the Spanish Broom and is in flower at the same time. Cut it hard back in spring.

Leycesteria formosa. This shrub of straight, cane-like shoots and purple and white tassels, followed by purple berries, does well on rather poor, sandy soil, growing to about 5 or 6 feet. It is not averse to lime. Prune hard in spring.

Ligustrum ionandrum. This is more reliable on poor, dry sandy soil than lonicera, keeping better furnished with dainty oval leaves. It does not rob the ground in the same way as the common privet, and does better on acid sandy soil than a limy soil. *Ligustrum ovalifolium aureum*, the Golden Privet, is a striking and beautiful shrub when grown as a bush but needs a good deal of moisture.

Lion's Tail (See LEONOTIS).

Lonicera (Honeysuckle). One of the very best of the climbing honey-suckles is the evergreen *Lonicera japonica halliana*, with profusions of creamy-white, very sweetly scented flowers from May until autumn. No situation or soil comes amiss to this robust climber. Others worth trying on poor, sandy soils are *L. giraldii, L. japonica aureo-reticulata, L. j. flexuosa, L. similis delavayi* and *L. purpusii*.

Lupinus arboreus (Tree Lupin). This is one of the fastest growing shrubs for excessively sandy soil; it will grow on a heap of sand if it has been deeply disturbed. Unlike the herbaceous lupin, it has no dislike of lime and is entirely at home on coastal sands in extreme exposure to wind and salt. Plants are not long lived—three years is about the limit—but they are easily replaced from seed. The colours are white, mauve and a really beautiful sunny yellow. It reaches a height of about 6 feet.

Lycium (Tea Tree). *Lycium chinense* is a rambling shrub with small pink flowers which are followed by scarlet fruits. It will grow in pure sand and may be used for fixing an arid sandy bank.

Magnolia. Members of this magnificent family should not be excluded from a free sandy soil of open texture, though a few need a peaty soil, which is sometimes found overlying light acid sand. Peat or decayed leaves may be added with advantage when planting, for the young plants to root into, and since they are surface-rooting and resentful of deep cultivation, surface mulches are extremely beneficial each spring before dry weather sets in, to keep the roots cool and moist. Since magnolias are not cheap to buy, expert advice from a local source should be sought as to which species are likely to be successful, as many are not lime-lovers.

Mahonia. Although the beautiful winter-flowering *Mahonia japonica* finds a sandy soil too thin and lacking in nourishment, no soil is too poor for the useful *M. aquifolium*. It thrives anywhere, on a sandy bank in full sun or in the deep shade of trees. With the approach of winter the handsome foliage takes on some shade of red or purple, while in starvation conditions of arid soil it often turns a bright red, which is almost scarlet. Though the vigorous masses of handsome leaves are one of the joys of winter, in spring they become again a bright polished green, and the bushes are gay with bunches of erect yellow flowers beloved by bees, later to become loaded with clustered masses of blue-purple berries. It is such an easy shrub and so shamefully neglected in modern gardens. A tall-growing variety of it is *M. a. undulata*, with glossy leaves, slightly waved at the edges, and bright yellow scented flowers, borne with great prodigality in early spring. It will reach 6 feet on sandy soil.

Manuka (See LEPTOSPERMUM).

Medicago. *Medicago arborea*, the Moon Trefoil, is an extremely fast-growing, loose shrub of up to 6 feet, with grey-green leaves and clusters of small, orange-yellow pea-flowers for most of the year. It is not a choice shrub, though it is useful for sandy wastes close to the sea.

Mexican Bodkin Bush (See COLLETIA).

Mexican Incense Bush (See EUPATORIUM).

Mexican Orange Blossom (See CHOISYA).

Mimosa (See ACACIA).

Moon Trefoil (See MEDICAGO).

New Zealand Flax (See PHORMIUM).

New Zealand Tea Tree (See LEPTOSPERMUM).

Olearia. The Daisy Bushes from New Zealand include excellent shrubs for light sandy soils in mild inland districts or in maritime counties of the south and west. Their leathery leaves make them among the toughest and most salt-tolerant evergreens for sandy foreshore gardens. The hardy *O. haastii* is a compact bush of oval grey-green leaves and off-white daisies in July and August. The wind-hardy *O. albida* is a taller bush about 8 feet high, with green privet-shaped leaves which are lost beneath a mass of rounded corymbs of off-white daisies a little later. This species is a splendid tough evergreen for an exposed position on poor soil.

Olearia macrodonta, with grey-green, holly-like leaves, makes a fine hedge in mild localities. Sandy soil should be enriched with decayed compost at planting time, and since young plants may suffer damage from exposure to cold winds, spring planting is best. The larger-leaved *O. macrodonta major* makes a fine, rounded bush, smothered with musk-scented white corymbs of flowers in June. Flowering is most abundant in sheltered, frost-free districts or by the sea. Large pieces strike readily in sandy soil. The heath-like *O. solandri* has golden, thyme-like, small leaves and golden stems, striking a bright gold effect among shrubs of darker foliage. It is sometimes found wild on sand-dunes. Where a sandy soil is enriched other members of the family will succeed. For *O. traversii* see Chapter 9 on Trees.

Osmarea × burkwoodii. A first-class evergreen with small dark green, pointed leaves and tiny scented white flowers in May. It is sufficiently hardy for the southern half of the country, and though it is reputed to prefer an acid soil, is doing remarkably well on our calcareous sand.

Pachysandra terminalis. This is a good, carpeting, semi-woody plant with bright green foliage for shady situations on light dry soil. It reaches a height of up to 10 inches.

Paeonia lutea Ludlowi. Though a sandy soil is often too thin and poor for herbaceous peonies, this Tree Peony has done extremely well on our sandy soil. It does not mind the lime in the sand, though it is equally at home on an acid sand. The strong, straight branches fan out from the ground to carry very handsome deeply-cut foliage. It is worth growing solely as a foliage shrub, for the butter-yellow, single flowers, though numerous, have too short a life. Our own plants, now 5 feet high, were a mass of flower five years after being raised from seed. Plant with protection from cold east winds.

Perovskia atriplicifolia (The Afghan Sage). This soft-wooded shrub, 3 to 4 feet tall, is a beautiful sight on limy sand. It is reasonably hardy, an ardent lime- and sun-lover, flowering best on thoroughly quick-draining soil. The slender whiplash wands are a chalky-white and the foliage a silver-green, while the flowers in late summer are a soft violet-blue. Cut it hard in the spring.

Philadelphus. These beautiful flowering shrubs are valuable for their June and July flowering when the great blossoming of shrubs is past its best. They prefer a light, rather dry soil, and an acid sandy one to a calcareous one. Their sweetly scented white or creamy-white flowers are carried in great profusion if flowering shoots are pruned as soon as they have flowered. Good kinds for modern gardens are the dwarf and bushy Manteau d'Hermine, double white, and Sybille, whose dainty flowers have purple centres. Belle Etoile and Beauclerk are beautiful varieties, taller and with a wide spread.

Phlomis. A genus of grey-leaved shrubs allied to sage, with curious whorls of flowers on erect stems. *Phlomis fruticosa* is the Jerusalem Sage, a 4-foot bush of soft grey-green leaves and unique whorls of golden flowers fading to white at the base. *P. cashmeriana* has rose-pink flowers, felted stems and paler silver leaves. They are only fairly hardy, but reliable on very dry, sandy soils, especially on the coast. Trim back after they have flowered in June.

Phormium tenax (New Zealand Flax). The finest plants I know are growing on wet sand. From large crowns of erect, sword-like leaves rise arching spikes of dull coppery-red flowers in summer. It is an imposing 7- or 8-foot shrub of sub-tropical appearance, the typical species having broad sea-green leaves, while others may be purple or striped with creamy-yellow. Good for open spaces by the sea.

Phygelius capensis. The Cape Figwort is a true shrub in the milder counties though in colder areas it is often wholly herbaceous. I recall a plant, growing on pure gravel, which reached 20 feet up the eaves of a house, though 6 feet is more usual. It makes an attractive shrub against a grey or white wall, invaluable for the lateness of its hanging coral-red flowers, which persist into autumn. It also cuts well.

Piptanthus laburnifolius (*P. nepalensis*). The evergreen Bush Laburnum, a 5- to 6-foot shrub with handsome leaves and clusters of bright yellow pea-flowers borne stiffly erect in May and June, is not entirely hardy in the open garden except in mild localities, though the protection of a wall brings it through cold winters in colder districts. It is best suited by a soil on the dry side and is not averse to lime. Seeds ripen in quantity and afford the best means of propagation. Young plants should be grown in pots for setting out in their permanent quarters.

Pittosporum. The two most suited to very sandy soil are *Pittosporum crassifolium* and the hardier *P. ralphii*, with larger, flatter leaves. The latter is a tough shelter shrub for the south or west. Both are largely used for hedges in the Isles of Scilly.

Potentilla. The potentillas are exceptionally good for small gardens, flowering all summer, with attractive tiny leaves and open strawberry-like flowers. Though the flowering of a single plant may have a slightly spotty effect, a group of three closely planted is very effective. They grow and flower well even on poor, stony soils and are quite hardy, only asking a place in the sun.

The tall *Potentilla fruticosa vilmoriniana* has silver foliage to set off the ivory saucers, and grows bushily to 4 feet. Klondyke is rich gold and Primrose Beauty creamy-yellow. *P. arbuscula* is exceptionally good, a 3-foot bush of rich yellow flowers, while among the dwarfs I like *P. fruticosa beesii* with silver leaves and yellow flowers. Tangerine is apt to fade in bright sunlight, needing some shade and good garden compost added to the sand.

Potato Plant (See SOLANUM).

Pyracantha. We have come to think of the Firethorns as essentially wall shrubs, though they make highly successful bushes in the open garden where they flower and berry well. They all have white flowers reminiscent of the May, but it is the bright berries which are their chief attraction. The much grown *Pyracantha coccinea lalandei* has brilliant fruits of orange-red, which in the past have been stripped in places where birds were plentiful, while the crimson berries of the tall *P. atalantioides* (*gibbsii*) remained till March. Now birds may be deterred from devouring the berries by spraying the bushes with Morkit, as directed by the makers; it is obtainable from any good garden stores. *P. rogersiana aurantiaca* and *P. r. flava* are desirable shrubs for small gardens, and their yellow fruits a pleasant change from the scarlets and crimsons. All are hardy and easy of cultivation, seeding freely on our limy sand.

Rhus. The Sumachs are not difficult as to soil, rather enjoying poor, dry ones. The well-known *Rhus cotinus* is now named *Cotinus coggygria* and is described on p. 47. The Stag's Horn Sumach, *R. typhina*, is an attractive small foliage tree with ferny leaves which turn red and orange at the fall of the year.

Rock Rose (See CISTUS).

Romneya (Californian Tree Poppy). *Romneya coulteri* and *R. tricho-calyx* are first-rate shrubs for sandy soils and *R. hybrida*, a cross between them, appears to have inherited the best qualities of its parents, with a neater habit and flowers as beautiful as those of either. It is one of the delights in late summer of a garden on thin, hungry sand. It loves a warm, dry soil and all the sun it can get, and given these, will grow to about 6 feet. If cut to the ground each spring, vigorous new shoots will appear, clothed with elegant blue-green foliage and surmounted by enormous, white poppy-flowers, with petals like crumpled silk and a gleaming cluster of golden stamens at the centre. It is not easy at the start and it is really essential to obtain pot-grown plants to begin with, planting in spring with as little disturbance of the roots as possible, even to the extent of breaking the pot. It may take a year or two to get going but once it has made up its mind to grow for you there is no stopping it, only prolonged frost will harm it, and even then it will spring up again from one of the strong wandering roots.

Rose of Sharon (See HYPERICUM).

Rosemary (See ROSMARINUS).

Rosmarinus (Rosemary). Rosemaries should have a hot, sandy soil to flower really well. Do not be content with only the type, *Rosmarinus officinalis*, whose pale lavender flowers appear early in the year, but grow some of the more beautiful varieties with intensely blue flowers or even pink. These are more tender and need shelter from cold winds. *R. o. angustifolius* is a dense bush of fine, feathery foliage and bright blue flowers, a truly beautiful tall shrub. The Corsican form of prostrate rosemary spreads out to cover 4 feet of ground, and has the same intensely blue flowers. The correct name for the tender pale blue rosemary which may be seen clinging to sunny walls is *R. lavandulaceus*. It is a pretty shrub though easily killed. Majorca Pink makes a pleasing change among the blues; it is a tall erect plant whose stiff stems are covered with flowers of soft lilac-pink very early in the year. Severn Seas is a good bright blue and Tuscan Blue, if it can be induced to flower, is most distinct, with broad pale green leaves and china-blue flowers.

The pretty gilded rosemary is still uncommon and we get many enquiries for this gay-foliaged shrub; it came to us from Corsica and is kept going from cuttings each summer. It is most beautiful when lit by the winter's sun. The flowers are a pale lavender. For an aromatic hedge, plant Miss Jessop's Variety, *R. officinalis fastigiatus*. It is undoubtedly the best and hardiest for the purpose, since its growth is strong and sturdy, quickly reaching the desired 3 feet.

Ruscus (Butcher's Broom). Our native Butcher's Broom, *Ruscus* *aculeatus*, faces up to almost impossible conditions of poverty of soil and drought. It is not an outstanding shrub, though its spiny, green form and bright red berries, the size of a pea, which are produced in great profusion if male and female plants are grown, and its endurance under the shade and drip of trees, make it a useful, low-growing shrub for difficult places.

Ruta (Rue). *Ruta graveolens* Jackman's Blue, with or without the yellow flowers at midsummer, is a worth-while 3-foot bush with blue-green leaves, which provide a delightful contrast for shrubs with purple or silver foliage. Trim over the plants in early spring.

Sage (See SALVIA).

Sage, Afghan (See PEROVSKIA).

Sage, Jerusalem (See PHLOMIS).

Salt Tree (See HALIMODENDRON).

Salvia (Sage). The aromatic sages are good plants for sandy soils and sunny situations. Our common sage, *Salvia officinalis*, is itself a pretty shrub with soft grey-green leaves and purple flower spikes in June. Some with variegated leaves, notably the gold-margined, and the pastel-shaded purple *S. o. purpurascens*, are delightful. *S. grahamii* makes no great show, but the pretty cherry-red flowers keep coming till late autumn in a sunny spot, and the tall, rather tender *S. involucrata bethellii*, with stout stems, large leaves and flowers of rosy-red, is a useful late-flowerer.

Santolina (Lavender Cotton). No soil is too dry and sandy for the santolinas and they thrive exposed to salt-laden winds. They are a most useful race of hardy shrubs with distinctive grey or silvery foliage for furnishing dry walls or impoverished sunny slopes. We grow three distinct types in our calcareous sand: the feathery *S. neapolitana* which makes a beautiful silver-leaved specimen, *S. chamaecyparissus* with stiffer frosted-white foliage, and *S. virens*, whose foliage is a rich bright green instead of silver. All have button flowers of gold or some shade of yellow in July and reach a height of about 2 feet. Cut back almost to the base of the previous year's growth each spring.

Sea Buckthorn (See HIPPOPHAË).

Senecio. This is a genus containing many distinct and valuable shrubs for gardens on sand. Some are not frost-hardy, though they are good sea-side plants. *Senecio laxifolius* is a foliage shrub of the first order, whose silvery-green leaves create a pattern of beauty quite apart from the blaze of colour of the yellow daisies in June.

Senecio monroi is another very desirable shrub with crinkled leaves silvery-white underneath, and yellow daisies coming into flower a couple of weeks after *S. laxifolius*. This proves extremely wind-resistant in exposed situations, and thrives in a dry, arid position. *S. glastifolius* from South Africa is a tender, rather rampant shrub whose tall stems carry branching sprays of lilac-pink daisies from April to June. It is difficult to keep it through the winter but cuttings are simple and should be planted out into the garden as soon as danger of frost is past. *Senecio cineraria* is an attractive 2-foot bush of silver, deeply-cut leaves; it is not reliably hardy except by the sea, where it seeds itself freely, but it is easy from cuttings or from seed. The superb White Diamond is a first-rate dwarf foliage shrub of more solid foliage, which is altogether whiter. *S. leucostachys* has finely-divided pinnate leaves and slender shoots clothed in silvery-white. Dry soil and full sun suit them all.

Shrubby Germander (See TEUCRIUM).

Smoke Plant (See COTINUS).

Snowberry (See SYMPHORICARPOS).

Solanum (Potato Plant). *Solanum crispum* begins flowering in April and continues without a break till autumn to clothe a sunny wall in mild localities with large clusters of pale mauve-blue flowers with yellow centres. The more meagre and stony the root run the better will it succeed; it needs nothing in the way of feeding. The poorest of sandy soils can be its home. It is a border-line climber susceptible to severe frost, though it does well trained against a sunny south- or west-facing wall, even in bleak localities.

South African Lion's Tail (See LEONOTIS).

Southernwood (See ARTEMISIA).

Spartium (See Chapter 7).

Spiraea. *Spiraea* × *vanhouttei* is one of the best spring spiraeas for poor soil. The pure white flowers wreath the arching branches in May and it has pretty autumn tints. It makes a good tall flowering hedge.

Spotted Laurel (See AUCUBA).

Stag's Horn Sumach (See RHUS).

Strawberry Tree (See Chapter 9).

Sumach (See RHUS).

Symphoricarpos (Snowberry). The common white Snowberry, *Symphoricarpos albus laevigatus* is too invasive for the ordinary garden though there are others with attractive berries for winter decoration, with less inclination to wander. White Hedge is a nice compact one with good white 'moth-ball' berries. Constance Spry is also good but a bit too invasive for the small garden. Mother of Pearl has berries that look pink but are really white stippled with crimson, and Magic Berry is a real crimson. They are excellent shrubs for a poor, sandy soil and are not averse to lime. They can reach a height of about 5 or 6 feet. All are eagerly sought after by flower arrangers.

Tamarisk (See TAMARIX).

Tamarix. We have come to regard the tamarisk as a seaside shrub or small tree where it grows on pure sand and is impervious to salt or sand blast. It is equally at home in inland gardens where trouble is taken to produce a beautiful flowering specimen with delicate feathery foliage and rosy-pink plumes of flowers. *T. pentandra* flowers at the end of summer and should be hard pruned in spring. The best of the spring flowerers is *T. parviflora*. The whole bush is smothered in April with small reddish-purple buds opening to flowers of deep pink. Prune it as soon as it has flowered. New plants are made by pushing pieces of last year's wood, a foot long, into sandy soil in February.

Tea Tree (See LEPTOSPERMUM).

Teucrium (Shrubby Germander). *T. fruticans* is a pretty shrub for warm coastal gardens on sand, succeeding even in extreme exposure. It grows rapidly to a 6-foot bush of long arching stems clothed in white, white-backed leaves and pale lavender flowers. In colder localities it is well worth planting against a wall. The very beautiful *T. f. azureum* is more tender, with flowers of the deepest blue and should have a sheltering wall in a warm garden. Both are easy from cuttings.

Thymus nitidus. All the thymes seem to love a warm, sandy soil and the shrubby *Thymus nitidus* is no exception. I have grown it in poverty-stricken conditions of poor soil and been repaid by a small bush with aromatic grey foliage and soft lilac flowers. It likes a hot, sunny situation.

Tree Lupin (See LUPINUS).

Tree Mallow (See LAVATERA).

Tree Peony (See PAEONIA).

Tree Poppy (See ROMNEYA).

Tree Purslane (See ATRIPLEX).

Ulex (Gorse) (See Chapter 7).

Veronica (See HEBE).

Viburnum. Most of the viburnums are woodland plants and need a richer soil than one on sand, unless it is enriched with leafmould and decayed vegetable matter, but that great gardener, the late A. T. Johnson, used to say that the old laurustinus, *V. tinus*, was the best drought-resister he ever grew on his light, hungry soil. I have known it make a valuable addition to the winter garden on extremely sandy soil, but it needs some help at the start in the form of moisture-retaining humus. If it were not so hardy and so easily cultivated, we should prize it more for its lacy white flowers in mid-winter. It needs room to achieve its maximum height of 10 to 15 feet in time. There is a useful pink-budded form suitable for small gardens, slow-growing to 5 feet tall, by 3 feet wide. Viburnums love lime and *V.* × *burkwoodii* and *V. carlesii* do well on our calcareous sand.

Wisteria. This beautiful deciduous climber is of easy cultivation, long-lived and hardy in the southern half of England and, belonging to the *Leguminosae*, is very happy where the soil is sandy. Where it is limy add some good loam, and a mulch of spent hops in March is much appreciated. A sunny position facing south or west is best, and some kind of support, either a wall or pergola is essential. Its tremendous vigour is rather too much for the small house, but it can be very effectively trained as a standard in the open garden with its branches spread out umbrella-wise and supported all round, so that the ropes of pale lilac flowers at the end of May hang out in great beauty. There is, too, a beautiful white form. The best-known variety and the most suitable for walls is *W. sinensis* but as a specimen in the garden I prefer the Japanese *W. floribunda macrobotrys*. Failure to flower abundantly is sometimes due to insufficient pruning. Shorten the leafy shoots to about five buds from the base in early August and again in winter to three or four buds.

Yucca. These are striking plants, tremendously long-lived, with sword-like leaves and exotic flower-spikes of creamy bells. Given a sunny position—they never flowered better than during the hot summer of 1959—they thrive almost anywhere, even on sand-dunes exposed to salt-laden gales. The gorgeous *Yucca gloriosa* takes time to recover after giving a superb display with magnificent spikes of creamy flowers tinged with red or purple, though the hardy and free-flowering *Y. filamentosa* flowers annually, in a profusion of hanging yellowish-white bells in July and August from a low tuft of glaucous leaves. It spreads slowly and surely by side growths which may be used to increase the stock. This yucca flowers at an early age.

17 The bearded irises provide a magnificent display if massed together in an open, sunny border

18 *Below: Kniphofia* Atlanta bears its red and yellow 'pokers' early in the season, and, as can be seen, does well on sandy soil

19 *Left:* The modern lupin hybrids are spectacular in their colouring and are perfectly happy on acid sand

20 *Right: Verbascum* Mont Blanc, with white flowers and silvery-green leaves, is an effective plant for lightening dark places

21 *Left:* Cynthia is a variety of rhododendron with particularly lovely crimson-red flowers. It is slow growing to a height of about 10 feet

CHAPTER FIVE

Perennial Plants

SINCE the main purpose of a garden is to give pleasure to its owner or to those who visit it, there is no surer way to achieve this objective than to fill it with plants which enjoy the prevailing climatic and soil conditions.

The strong line of demarcation between acid and alkaline sand, to which I have referred, is not the barrier to hardy herbaceous plants that it is to many shrubs. Certain plants, such as scabious, dianthus, iris and aubrieta prefer some lime, while others, like lupins, have a preference for an acid soil. By and large, however, perennial plants do not object to lime in the soil; they are more concerned with drainage, whether the soil is light or heavy, and if it has sufficient nourishment for their needs, for some sicken with rich fare and others pine without it.

Curiously enough, people who garden on light, sandy soils seem to want to grow plants which like damp and then are surprised that they refuse to grow well, so that in sandy districts one often comes across arid sandy wastes which might be furnished with plants more tolerant of such conditions. To get the best results from light, sandy soils there are the following possible alternatives. Either one can incorporate into the soil sufficient humus to transform the actual character of the sand and make good its water-holding capacity—a slow process which takes years to accomplish—or one can restrict oneself to the non-thirsty, drought-resisting plants which thrive naturally on sandy soils.

In either case the preparation of the ground should be as thorough as one can make it, with generous applications of bulky organic material to bind the sand and to act as a sponge to retain moisture, and since sandy soils are normally deficient in potash this should be added either as part of a mixed fertiliser such as Growmore, or as sulphate of potash.

Where lack of labour or materials make extensive preparation of the ground an impossibility, fairly satisfactory results can be achieved by limiting the classes of plants to those known to succeed in a dry, sandy medium, and by laying a surface mulch of organic matter around the plants during the hottest months to keep the roots cool and moist. Choose a day after rain and lay it not deeper than a couple of inches.

When and how to plant. On light sandy soil it is safe to plant at almost any time from September to April in open weather. In spring, however, drought and drying winds are very destructive to newly planted subjects, and where spring planting must be carried out, the plants should be puddled in, mixing the soil with water in a bucket and dipping the plant into the mixture before planting. Firm planting is essential on shifting sand, fresh soil being raked around the plant afterwards.

Which plants are best. Since at the outset our sandy garden was extremely low in nutrients we were forced to make a special study of those plants which do well with the least nourishment and do not care for an excess of water. Summer is the time to judge which plants are least affected by drought. It is useless to rely on watering to keep them alive, for plants which require a good deal of moisture wage a losing battle on hot, dry sand, unless there is a water-table not far below the surface. Many sand-lovers come from hotter countries than ours and are at home on dry, sharply-drained soils. They should be a first choice for the garden on sand. We look for them among plants with waxed or woolly leaves, with white or silvery foliage, or those whose leaf surfaces have been so reduced they are little more than ribs or spines. This, in a very general way, is a guide to those plants which will succeed on sand.

Achillea. Few families are more useful for furnishing light soils with a low rainfall than the yarrows. *Achillea filipendulina* Gold Plate is a tall plant with stout, woody stems and flat, yellow flower-heads often 6 inches across. It is good for the back of the border or the centre of an island bed. *A.* Coronation Gold is similar, though only half the height. *A. clypeolata* and *A. taygetea* are both good, with silvery, finely-divided leaves, but neither is as reliably permanent as the hardy Moonshine, whose flat heads of flower in May and June are a bright unfading yellow above delicious tufts of greyish leaves. I would grow it for its leaves alone. *Achillea ptarmica* The Pearl is in demand as a cut flower for its small white buttons on branching stems, but is apt to run in the border.

Alstroemeria. The orange-yellow *Alstroemeria aurantiaca* is the hardiest, while the more tender *ligtu* hybrids are in soft pastel shades of salmon-pink and lilac-rose. Plants go on for years once they are established. The alstroemerias go deep—7 to 9 inches is not unusual—but the fleshy, brittle roots break at the slightest touch and the easiest way to start a colony is from seed. This is simple, for the seed is large and can be spaced out an inch apart, three seeds to a 3-inch pot, in May or June. When the seedlings are a few inches high, tap out the whole contents of the pot, without disturbing the roots, into a prepared bed, planting each clump a foot apart and 6 inches deep. Later they will find their own level.

Lovely as the alstroemerias are, they can be a nuisance in the mixed border where their invasive tendencies cannot be checked, and they are better in a bed to themselves. Once they are established they seed freely, flinging their seeds about with abandon so that plants appear in the most unlikely places.

The *ligtu* hybrids like a sunnier, more open site than the type, the parents species coming from drier areas of Chile. All are excellent cut flowers, lasting quite three weeks in water.

Anaphalis. Useful grey or silver-leaved plants for poor, dry, sandy soil. *Anaphalis yedoensis* has pearly-white 'everlasting' flowers on upright 2-foot stems in late summer, and is pretty from the time its silvery shoots appear; its only fault is that it is inclined to run a little. Plant it where the chance of success seems small and you will be pleased with it. *A. triplinervis* is a 'stay at home', altogether neater and shorter and more woolly but with similar 'immortelle' white flowers. They can be dried for winter decoration.

Anchusa. Two good anchusas for light sandy soil are *Anchusa azurea* Loddon Royalist, an outstanding perennial with large blue flowers on 3-foot stems, and the compact *A. caespitosa*, with flowers of gentian-blue with a small white eye. They are in flower during May and June. Neither will put up with heavy soil or winter wet.

Anthemis. It is a pity that *Anthemis tinctoria* is not longer lived except on sharply-drained sand; newly planted pieces are so much freer with their yellow daisies than older plants with woody rootstocks. Detach small rooted pieces from the parent plant and replant in spring or autumn. The orange-yellow *A. sancti-johannis*, Grallagh Gold and Beauty of Grallagh may be enhanced by association with pale lemon varieties such as *kelwayi* or Mrs E. C. Buxton.

A. cupaniana is one of the joys of our garden in early spring, tumbling down a wall, clothing some unsightly corner or finding a home in the roughest grass. It is a splendid weed-smotherer and though a little free with its seeds on light soil, the dense, grey-green cushions of ferny foliage are gay with long-stemmed white daisies from April to June.

Arenaria (Sandwort). The common name of sandwort is an indication of the places the sandworts naturally inhabit, many species growing happily where sand accumulates above high-water mark in various parts of Britain. In the garden they delight in sandy soil, and though they like good drainage they do not care to be burnt up in hot summers. Most have small white flowers. The summer-flowering *Arenaria montana* has an almost trailing habit and *A. ledebouriana*, with green-gold leaves and pretty white stars, likes to hang down in pockets in the rock garden. *A. verna*, now *Minuartia verna*, is a cushiony hummock starred with tiny white flowers in spring. I find *A. balearica* is the best known; it loves a damp, shady situation, particularly where it can clamber over cool stones forming a dense carpet of minute green leaves studded with tiny white stars. This is a good plant for taking the new look from one of the modern plastic pools, for it enjoys the drip from a sprinkling fountain.

Armeria (Sea Pink or Thrift). *Armeria maritima*, the pale pink sea pink, has several garden varieties worth growing on very sandy soil. They range in colour from white, through rose-pink to deep red. The dwarf *A. m. alba* is white, Vindictive is rose-pink, and Ruby Glow and Bloodstone are taller and deeper in colour, good reds. The green, grassy tussocks are not easily divided as they spread out from one main root. Small pieces root readily in a sandy frame. For border work the best of the thrifts is Bees' Ruby with large, globular heads of rose-pink on tall stems. It is easy to propagate from seed.

Arnebia echioides. This is the Prophet Flower, since the legend has it that Mahomet touched the flower with the tips of his fingers, causing those dark brown spots which appear almost black at first against the bright yellow of the flowers. The arnebia belongs to the borage family and is perfectly content in a poor, dry soil in the rock garden where it flowers in April on 9-inch branching spikes.

Asphodeline lutea. A native of the Mediterranean regions, *Asphodeline lutea* is entirely at home on warm sand, with 3-foot stiff spikes of yellow, fragrant flowers opening at intervals, never many at the same time.

Aster (Michaelmas Daisy). The tall Novi-belgii and Novae-angliae types of michaelmas daisy are a waste of time on light sandy soils which tend to dry out during the summer. They are prone to mildew and lose their lower leaves. They need good loamy soil, but the varieties of *Aster amellus* do extremely well on sandy soil, providing useful colour from August to October. They have large flowers and most do not exceed 2½ feet in height. They seldom require staking and may remain in one position for years. Plant in spring when new shoots are seen appearing. You cannot do better than the violet-blue Bessie Chapman, the pink Sonia, the very robust Mauve Beauty, and the large rosy-lavender Nocturne.

Aubrieta. The dwarf *Aubrieta deltoidea*, with other rock garden plants such as cerastium, alyssum and arabis, is one of the highlights of the garden on sea sand but is difficult on acid sandy soil, preferring some lime.

Bergenia. Once called *Saxifraga*, then *Megasea*, this trouble-free plant accommodates itself to any soil, doing particularly well where there is lime. The huge fleshy leaves are outstanding for clothing odd corners, as edgings, or against stonework, and the flowers, usually pink, on stout foot-high stalks, come very early in the year. *Bergenia cordifolia* is very commonly grown; Evening Glow is more compact with spikes of reddish-purple flowers and reddish leaves. *B. delavayi* has huge leaves which turn beetroot-red in winter, remaining so until the spring. All lend themselves to division.

Borage (See BORAGO).

Borago (Borage). *Borago officinalis*, the common borage, and *B. laxiflora* are useful blue-flowered plants with coarse, hairy leaves for naturalising rough places on sandy soils. They seed too freely for sophisticated gardens.

Californian Fuchsia (See ZAUSCHNERIA).

Campanula. The range of campanulas for sandy soils is extremely wide, from the tall *Campanula lactiflora*, which likes to ramp, to the dwarf *C. portenschlagiana*.

Campanula carpatica is very good on light soil, with blue or white bells or salvers which remain open for many weeks. *C. glomerata dahurica* makes a gorgeous bit of rich purple colour and though it has a bad reputation as a spreader, I have never found it a nuisance on our poor soil. The rampant, trailing *C. poscharskyana* E. K. Toogood has pretty starry flowers of light mauve-blue and seeds freely. Most flower during June and July.

Cape Figwort (See PHYGELIUS).

Cardoon (See CYNARA).

Catananche. A splendid perennial on light sandy soil with good drainage, which is impervious to drought. *Catananche caerulea major* is altogether finer than the type. It must be propagated by root cuttings, whereas the type is raised easily from seed. Catananches need full sun to produce a succession of their blue, cornflower-type flowers on long stalks from June to late summer. The silvery bracts are an added attraction. There is also a pretty white form. At the end of August cut the plants back lightly, this will ensure them a longer life. They have long tap roots, so that the normal means of increase is from seed.

Catmint (See NEPETA).

Centaurea. Good plants for light soils are *Centaurea dealbata steenbergii*, with greyish leaves and large rose-purple flowers on 2-foot stalks, and the taller *C. macrocephala* which carries its huge yellow thistle-heads in June.

Centranthus (See KENTRANTHUS).

Ceratostigma plumbaginoides. A valuable carpeter for sandy soils with starry flowers of gentian-blue in early autumn and leaves which turn a pretty reddish-bronze before they are cut down for the winter.

Cheiranthus. Perennial wallflowers are none too common though it would be a loss were we to lose them altogether. The old-fashioned Harpur Crewe makes a sturdy bush, covered through May and June with masses of small, scented, double flowers of bright yellow. Moonlight has pale yellow flowers at the same time. They enjoy the poorest soil with some lime in it and a sunny place. Ideal for the top of a dry wall. Old plants should be replaced by young ones which root easily from cuttings taken at mid-summer.

Chrysanthemum. Though few outdoor chrysanthemums are a success on very sandy soil, *Chrysanthemum rubellum* is an exception. This trouble-free perennial grows bushily, and stands erect when grouped, as it always must be. The large, single flowers are soft chalky-rose, and associate beautifully with the mauves and lavenders of the varieties of *Aster amellus*. There are single and double forms in a variety of colours but none is better than the type. New plants are made by detaching small side pieces in the spring.

Convolvulus mauritanicus. This is a beautiful trailing plant to cherish in some warm spot on a light sandy soil. It is not so wholly to be trusted that we can ignore the precaution of taking cuttings in late summer and overwintering them in a frame. Plant it to hang down a wall in full sun, where the vase-shaped flowers of soft periwinkle blue will delight for many weeks until frost puts an end to the display.

Crambe cordifolia. This close relative of our common seakale, *Crambe maritima*, which is a native of sandy shores, puts one in mind of a giant gypsophila, with clouds of dainty white flowers on tall branching stems rising from a mass of green cabbagy leaves. It needs ample space to display itself and looks most attractive against a dark background. *C. orientalis* with grey leaves is also good.

Cranesbill (See GERANIUM).

Cynara. *Cynara cardunculus* is the Cardoon, a gigantic thistle with decorative silver leaves and spiny purple thistle-heads, and *C. scolymus* is the Globe Artichoke, very similar in effect in the August garden. Both are escapees from the kitchen garden, well worth a place in the flower border.

Day Lily (See HEMEROCALLIS).

Dianthus. Here are included rock garden pinks, long-flowering *Dianthus allwoodii* pinks and border carnations. Their basic requirements are good drainage, free access to a current of air, a sunny position and sufficient lime in the soil. Every sandy seashore garden should have its quota. Avoid using either peat or leafmould, though the addition of good garden compost around the plants has very satisfactory results. A sprinkling of bonemeal is enough in the way of fertiliser. Plant firmly but not too deeply and do not bury the stem. Planting may be carried out in autumn or spring. Apart from the beautiful scented flowers many have attractive blue-grey foliage.

Dictamnus albus (Burning Bush). This plant bears spikes of white or lilac flowers during June and July. On a dry soil it should not be expected to attain a height of more than 2 feet. It exudes an inflammable oil which may be ignited on a warm, still day without damage to the plant, hence the reason for its common name.

Dimorphotheca barbariae compacta. This South African plant, which bears lilac-pink daises with a buff reverse all summer and has lemon-scented leaves, has proved hardy in most parts of the country. It is a spreading mat of a plant for sandy soils in a sunny position. Trim it over in early spring if it is inclined to straggle.

Echinops (Globe Thistle). Most of the echinops are too rough and coarse for gardens where space is limited and only the best is good enough. They are deeply-rooting perennials with spiny leaves, woolly underneath, and thistle-heads of slaty-blue, starting into flower in June and continuing on into October. *Echinops ritro* Veitch's Blue, however, deserves its Award of Merit from the Royal Horticultural Society. It is a 3-foot plant with bright ball-shaped-flowers on rigid stalks which need no staking.

Erigeron. Erigerons delight in a sandy soil with good drainage, and their bright daisy flowers—resembling those of the *Aster amellus* varieties—in mauve, lilac-pink or dark purple, are produced very freely during June and July. The new varieties are outstanding for their profusion of flower, and their compact growth, seldom exceeding 2 feet, commends them to the smaller garden. Dwarf kinds like Felicity, Dignity and Prosperity are suitable for the front of a border. The individual flowers of varieties such as mauve Serenity or clear pink Vanity are excellent for cutting and Violetta, with double florets of bright lilac-mauve, is outstanding. Not only are they brighter and clearer in their colours but they flower for months on end. It is unfortunate that some tend to have somewhat lax stems. Cut over as soon as they have gone out of flower in early autumn. The best flowers are provided by young plants and it is worth raising cuttings every year or two. *E. glaucus*, mauve or pink with fleshy green leaves, is a dwarf carpeter for poor soils.

The little dainty *Erigeron mucronatus*, with small pink and white daisy flowers, is in bloom all summer, and though it is difficult to eradicate once it has taken possession, it is useful for the crevices of walls or between paving-stones where it is allowed to seed freely. *E. aurantiacus* is unique in having orange flowers and is most reliably perennial where the soil is light and sandy.

Eryngium (Sea Holly). No family of plants is more at home on pure sand than the eryngium. Our native sea holly, *Eryngium maritimum*, thrusts its woody rootstock to tremendous depths into shifting sand, and if you should encounter this maritime plant with spiny leaves and huge bracts surrounding the blue teazles, resist the temptation to remove its creeping root, often 8 feet long, to your sandy garden, and grow instead *E.* × *oliverianum*, *E. agavifolium* or one of the handsome garden forms like *E. planum* Dwarf Blue. *E. giganteum*, often 6 feet tall, with flowers which become bone-white as they ripen, is not a reliable perennial, flowering best in a dry summer, but is so beautiful it should not be passed over, since it often leaves behind some useful seedlings. Few perennial plants give a more delightfully unusual picture on sand in late summer, but they are bad transplanters and should be planted when small.

Euphorbia (Spurge). Many spurges do well on light sandy soil. The giant *Euphorbia wulfenii* makes an enormous bush of blue-green foliage on our sea sand though the exotic lime-green flowers in early spring are unreliable, appearing in great profusion one year and giving us a miss the next. The semi-prostrate *E. biglandulosa* has bright glaucous-blue leaves and yellow bract-like flowers in winter, while *E. myrsinites* is no less steely blue with yellow flowers on slender stems in spring. *E.*

sikkimensis is inclined to ramble, and though *E. griffithii* sprawls a little, the variety Fireglow may be forgiven for its flowers of intense burnt orange.

Evening Primrose (See OENOTHERA).

Everlasting Pea (See LATHYRUS).

Flax (See LINUM).

Fuchsia, Californian (See ZAUSCHNERIA).

Geranium (Cranesbill). Those gardening on almost pure sand should try the hardy geraniums. They really come into their own in some shady place on a sandy soil. *Geranium sanguineum*, with magenta saucers, is a bit of an encroacher though splendid ground cover for barren soil between shrubs. *G. endressii* A. T. Johnson and *G. e.* Wargrave Pink are excellent weed-smotherers with a long season of flower.

It is often suggested that *Geranium wallichianum* Buxton's Blue is a plant for a sheltered, shady situation, but I have seen it growing lustily on a sunny, sandy bank, smothered in saucer flowers of purplish-blue with a white eye, for many months in late summer. The long arms spread widely to cover the ground from quite a small rootstock, so that it is easy to give the plants a little assistance in the way of rotted compost in the spring. It is hardier grown in a light sand soil than in a richer medium.

Geranium ibericum is well known for its large purple flowers and big, decorative leaves; it grows with us with the minimum of care, and is a fine border plant. The blue *G. grandiflorum* is extremely tough while the newer Johnson's Blue is similar with mauve-blue flowers.

Glaucium (Horn Poppy). The Horn Poppy is particularly suited to over-light sandy soils and for the arid part of the wild garden. It has deeply cut silver leaves and a succession of poppy flowers, bright yellow in *Glaucium flavum* and flame-coloured in its variety *tricolor*, followed by long curving seed-vessels. Cut down at summer's end, when they increase in size from a woody rootstock.

Globe Artichoke (See CYNARA).

Globe Thistle (See ECHINOPS).

Golden Rod (See SOLIDAGO).

Gypsophila. The roots of this plant go deep in search of moisture into a limy sand which it loves. It is a chalk- and lime-loving plant. The clouds of dainty flowers, white or pink, on branching stems, are delightful at the front of the border or in the rock garden. Bristol Fairy has double white flowers on wiry stems and is excellent for cutting. Rosy Veil has dainty, double pink flowers which it bears in great profusion and it has a semi-prostrate habit. The small *Gypsophila paniculata*, of which the foregoing are varieties, is a film of cloudy white, ideal for decoration with scabious or sweet peas or other cut flowers.

Hemerocallis (Day Lily). Recently introduced varieties of the Day Lily are in many delightful new shades of pink, orange-red, maroon and wine-purple as well as golden-yellow. In their modern dress they rank among the no-trouble hardy perennials which may be left undisturbed for many years. Though they are much used in border work I prefer them on their own, since they have very decorative green leaves to offset the succession of lily-like flowers in July and August. Try the deep pink Lady Ormerod, Crimson Glory, the lemon Vespers, the ruby-purple Black Magic and the cardinal-red Red Torch. Flowers of these newer varieties last more than the proverbial day. On very sandy soil start them off with generous helpings of humus and keep them watered in dry weather.

Houseleek (See SEMPERVIVUM).

Iberis (Perennial Candytuft). A shrubby plant smothered in May and June with rounded, snow-white heads of flower.

Iris. Though such irises as *Iris kaempferi* need moisture, the tall bearded iris, *I. germanica*, is tolerant of poor soil conditions with good drainage and grows to perfection if the sandy soil is enriched with humus and bonemeal. Indeed vegetable compost is the key to success on light soils, giving body and making the soil more retentive. Under such conditions the tall bearded iris and the small *I. pumila* do extremely well. They are sun-lovers and prefer an open position. Transplant every few years soon after they have finished flowering in July and very lightly cover the rhizomes on sandy soil.

I never tire of singing the praises of *I. unguicularis* (*stylosa*) on hungry sand. No plant better repays a starvation diet on poor soil wth sharp drainage. The greatest number of the delicate lavender flowers are produced in a sun baked situation on barren soil. Rich soil produces foliage at the expense of flower. Transplant only when necessary in April and lift in large clumps or the plants may fail to flower for a year or two. Though *I. sibirica* and its varieties are often regarded as moisture-lovers, they do well in very varied conditions of soil and are not averse to a light sand.

Kentranthus (Valerian). You may shudder at the thought of introducing the valerian into your garden, for it can be a menace from self-sown seedlings. It is, however, a useful plant for very dry barren places where little else can be induced to grow, and in its deep red form, *Kentranthus* *ruber atrococcineus*, should not be despised for its willingness to clothe walls of the most desiccated description. Its passion for seeding may be checked by the timely removal of faded flowers, when it will give a second crop of flower in early autumn. Kentranthuses are also listed under the name Centranthus.

Kniphofia (Red Hot Poker). The kniphofias rarely come to harm on light, sandy soil which has been deeply dug and manured, though cold, wet soils are often fatal. They are gross feeders but have been long-lived on our sea sand, when given generous helpings of humus at planting-time, and surface dressings spread around their roots in spring. They have good foliage, and this and the stateliness of their tall flower-spikes calls for a position of isolation away from other plants. Though the indestructibly hardy Royal Standard, the late-flowering coral-red Mount Etna and Yellow Hammer look best in broad clumps, for border work I prefer the species *Kniphofia galpinii* with thin rushy leaves and pokers of vivid flame, the ivory-white Maid of Orleans and midgets such as Gold Else and Bee's Lemon, though these last are not so reliably hardy.

Kniphofia caulescens is distinct, with a number of orange-headed pokers springing from a woody stem, a good plant for a show piece in a dryish situation in a seashore garden. The variety Atlanta is an early flowerer with broad, fleshy leaves and spikes of red and yellow, particularly good for very sandy soil in exposed places.

Lamb's Ear (See STACHYS).

Lathyrus (Everlasting Pea). *Lathryus latifolius* White Pearl, white flushed with rose, is a beautiful and rampant climber for poor sand soils. This easy, indestructible perennial survives in spite of neglect. Use it to cover an unsightly corner or to mantle a fence, where it will flower through July and August.

Lavender, Sea (See LIMONIUM).

Limonium (Sea Lavender). *Limonium vulgare* is common along the east coast of England where its purplish-blue flowers on leafless stalks are a magnificent sight from late June. Garden varieties such as Chilwell Beauty and Blue Cloud, with larger flowers, are splendid hardy plants for sandy gardens. Give them room for an eventual 2-foot spread.

Linaria (Toadflax). Two toadflaxes you may admit with pleasure to a sandy garden are *L. vulgaris*, with yellow, orange-flecked flowers on 18-inch stems, and *L. purpurea* whose tall stems decked with purple flowers are garden-worthy on excessively poor sandy soils.

Linum (Flax). All flaxes love a dry, poor soil. *Linum narbonnense* Six

Hills throws up a host of wiry stems, each bearing clusters of buds which open to give a dazzling show of blue from June to September. *L. perenne* has paler, slightly smaller flowers and scatters its seed far and wide over a sandy garden. Flaxes dislike division but are easily increased from seed. The slightly more tender *L. flavum* is a compact small plant with greyish leaves and bright yellow flowers on 9-inch stems. Good for the rock garden on light soil.

London Pride (See SAXIFRAGA).

Lupin. Lupins can carry on in the most arid of sandy soils if it is acid, but they dislike lime. In California they often cover burning, shifting sands in such profusion that they are used in desert reclamation or for fodder crops in rainless areas.

Macleaya cordata (syn. *Bocconia cordata*). The Plume Poppy is a tall, imposing plant which grows vigorously on sandy soil and may have to be kept in check, since it spreads by underground stems. Plant it where there is ample space for the large fig-type leaves, bronze-green above and grey beneath, and the delicate feathery plumes of small orange-buff flowers carried aloft on 6-foot stems in July.

Mallow (See MALVA).

Malva (Mallow). The herbaceous mallows are best in hot, dry soil. Two of the most desirable are the pink-salvered *Malva alcea fastigiata*, a 2-foot plant which blooms from June to August and *M. sylvestris* Primley Blue which spreads out to cover a considerable space with the lax shoots starred with veined blue cups. Rarely found in catalogues, this slate-blue variety is increased by cuttings, since it appears to be completely sterile. Plants are still found at Paignton Zoo, though the garden at Primley no longer exists.

Michaelmas Daisy (See ASTER).

Muehlenbeckia axillaris. This dense and charming carpeter of minute, heart-shaped leaves on wiry stems has many uses for frost-free gardens. It is impervious to salt-winds and may be used as a thick screen, clothing wire-netting or slatted fences in exposed places on the coast. It is an excellent carpeter through which small bulbs will grow and is nowhere more happy than mantling walls and hedges in the Isles of Scilly.

Nepeta (Catmint). Few perennial plants furnish a garden on sand with as long a season of flower as the catmint. The dwarf *Nepeta × faassenii* (*N. mussinii*) is useful for edgings, though the taller and finer Six Hills Giant has superseded it in modern gardens. Catmint is at home on any sand and a 'standing dish' for gardens by the sea shore. Let the top growth remain on the plant throughout the winter as protection against frost and trim over in early spring.

Oenothera (Evening Primrose). Since *Oenothera biennis*, the true evening primrose, may be seen yellowing our sand dunes in great profusion, we shall expect some of the perennials to grow happily in sandy gardens. The deep-rooting *O. missouriensis* is happy where conditions are poor and dry and exposed to full sun. This lovely plant grows close to the ground, throwing out long shoots with narrow, pointed leaves and pointed red buds which open to display the huge yellow salvers from June to well into September. *O. acaulis* rarely fails to attract attention by reason of the translucent quality of its large white cups. It is truly perennial only on well-drained sandy soil but not under wet conditions. Many of the perennial oenotheras with flat basal leaves seem more difficult to establish on sand that is hot and dry.

Onopordon (Scotch Thistle). *Onopordon acanthium* grows to 6 feet tall with immense branching white stems and large-spined grey leaves. The small rose-purple thistles are of secondary importance. It is easy to raise from seed.

Papaver (Poppy). You are either an admirer of the large Oriental Poppy, *Papaver orientale*, with its brief blaze of colour in early summer, or you will not put up with the untidy mess it makes after it has flowered, at any price. There is no disputing, however, that the poppies provide one of the brightest patches of colour on sandy soil at this time. Try some of the modern varieties in truly astonishing colours: Watermelon, cherry-rose; May Queen, dwarf, double and poppy-red; Mrs George Stobart, salmon-pink; or Indian Chief, mahogany.

Phlox subulata. Though the brilliant herbaceous phloxes prefer cool, moist conditions, the prostrate, carpeting Moss Phlox, *Phlox subulata*, from the sandy hills on the Atlantic side of North America, is entirely happy on a well-drained sand that is not burnt up in summer heat. Its great enemy is winter wet, and unless drainage is good, this densely tufted plant is liable to die. The evergreen mats, so attractive in the winter garden, are completely smothered with flower in late May and early June. The flowers are lilac, lavender or rose, often with a darker eye and there is a delightful white form.

Phygelius capensis (Cape Figwort). This South African plant has survived over 20 degrees of frost in the open in our garden. It is most often seen as a herbaceous plant where it is apt to run a little, but I like to grow it as a wall shrub, where its bright red, tubular flowers hang out from the rich dark foliage in profusion from July to September.

Plume Poppy (See MACLEAYA).

Polygonum affine. One recommends one of the invasive knotweeds with diffidence, but *Polygonum affine* and its varieties are wonderful garden plants with a long season of flower and no tendency to spread. *P. affine* flowers from early June into October and even November, when the green mat of leaves turns a delightful russet-brown. Its pink spikes are a pleasing contrast to the deeper flowers of Darjeeling Red and Lowndes Variety, which is even better, with large, dense spikes of deep red. These polygonums are ideal plants for the edge of a path or wherever low-growing plants are needed.

Prophet Flower (See ARNEBIA).

Red Hot Poker (See KNIPHOFIA).

Salvia. Most sages can be recommended for light sandy soil. *Salvia ×
superba* and its dwarfer and bushier variety Lubeca, and the still dwarfer East Friesland, are among the best hardy perennials we have, bearing their spikes of violet flowers tinged with red for quite two months during July and August; their flowering season may be prolonged if the faded flowers are cut off to encourage a second crop. *S. sclarea* Vatican is less truly perennial but seldom dies without leaving behind some of its progeny. A group of these plants with enormous grey leaves and tall spires of soft mauve-white achieved by self-sown seed is outstandingly beautiful.

Salvia haematodes, though better treated as a biennial, has come through normal winters in a very dry situation in our sandy garden, and its branching spikes of light mauve-blue are beautiful.

Sandwort (See ARENARIA).

Saxifraga umbrosa (London Pride). This plant, beloved of the Victorians, does well even on light soils when given partial shade. Its pale pink flowers spring from leafy basal rosettes, and there is also a form with variegated leaves.

Sea Holly (See ERYNGIUM).

Sea Lavender (See LIMONIUM).

Sea Pink (See ARMERIA).

Sedum. Sedums are specially suited by stony, sandy soil. *Sedum spectabile* is a first-rate plant whose glaucous leaves are attractive before the flat, pink flower-heads brighten the autumn garden. They flourish on any well-drained soil but are specially suited to seaside gardens on sand, their succulent leaves acting as storehouses of water, impervious to drying salt winds. Carmen and Meteor have flat heads of carmine-pink, and Autumn Joy changes from salmon-pink to coppery-rose as the flowers fade. The butterflies love them all.

Sempervivum (Houseleek). The common houseleek, *Sempervivum tectorum*, gets its name from its ability to grow on the roofs of buildings, and most of the sempervivums are happiest when they can spread their stiff rosettes on a barren spot in the sun. They are curious rather than beautiful and are grown primarily for their rosettes of green and purplish leaves, though they produce stiff little flower spikes of pink or reddish flowers in summer. All respond to division.

Sisyrinchium. The small *Sisyrinchium bermudiana* has iris-like tufts of foliage and dainty flowers of violet-blue on 9-inch stalks from April to August. Pole Star is taller but similar in habit with spikes of white flowers.

Solidago (Golden Rod). The old Golden Rod can be a menace on sandy soil, requiring drastic treatment to keep it in check, but newer varieties with pretty mimosa flowers on shorter stems are not inclined to run about, and deserve a place. Their bushy, floriferous growth will clothe rough places or is useful among shrubs. All are hardy perennials, quickly forming large clumps which lend themselves to division. They impoverish the soil and need replanting every few years. Good modern kinds are Mimosa, Peter Pan, Leraft and Lemore with yellow plumes in late summer, without the tendency to roam that spoils the older ones. The pigmy Golden Thumb, which I admired in company with the succulent sedums, is a delightful small plant, no more than 12 inches tall, with clouds of soft golden flowers in August and September.

Spurge (See EUPHORBIA).

Stachys lanata (Lamb's Ears). A fine foliage plant for poor, dry conditions. Its soft silvery leaves and mauve spikes of woolly flowers make it a good front-of-the-border plant.

Stokesia laevis. Do not miss the spectacular *Stokesia laevis* Blue Star which has attracted much attention in the island beds at the Royal Horticultural Society's Garden at Wisley. The large, shaggy, cornflower-type flowers of lavender-blue on a low-growing bush with dark, leathery leaves last for many summer weeks. Plant in spring.

Teucrium chamaedrys (Wall Germander). A dwarf, neat bush of dark green leaves and small spires of rose-pink in August.

Thrift (See ARMERIA).

Thyme (See THYMUS).

Thymus (Thyme). All thymes are lovers of light, sandy soil with good drainage and full sun. They include those used for culinary purposes, the common thyme, *Thymus vulgaris*, of which there is a pretty variety worth growing for its pink flowers, shrubby species, and creepers. Two favourites with me are the golden thyme, *T. × citriodorus aureus*, and the silvery thyme, *T. × citriodorus* Silver Queen, both sweetly scented. The prostrate wild purple thyme, *T. serpyllum*, is not bettered, in my opinion, by any of the garden forms, but these are delightful additions in crimson, mauve and white. All are useful for planting among the paving stones on terraces, since they may be walked over with impunity. The creeping thymes also make a beautiful thyme lawn, which remains green when grass on sandy soil has browned, and is a lovely sight when the purple or crimson flowers carpet it during the summer. The thymes have the advantage of preferring a poor sandy soil to a rich one and will cheerfully grow where greedier plants will fail.

Valerian (See KENTRANTHUS).

Verbascum. Though not among the long-lived perennials, many are extremely decorative, with imposing, tall spires which provide contrast in a border. They prefer a sharply-drained, light soil and I have seen them growing to perfection in very poor conditions.

Verbascum phoeniceum is in shades of lilac and purple, a stout branching plant on 2-foot stems. The tall *V. olympicum* has huge silver-felted leaves and 5-foot spires of yellow flowers, and though it is slow to reach flowering age, the leaves are highly decorative in the meantime. Some of the most beautiful are to be found among the hybrids —Cotswold Beauty, biscuit and lilac; Mont Blanc, white; Cotswold Gem, terra-cotta pink; and Pink Domino with dark leaves and deep lilac flowers. Seed is the normal means of increase.

Wallflower (See CHEIRANTHUS).

Wall Germander (See TEUCRIUM).

Zauschneria. The Californian Fuchsia, *Zauschneria californica*, is one of the most brilliantly coloured rock garden plants we can plant for autumn colour. It is a plant that loves to be baked by the hottest sun and in almost pure sand, which has been deeply disturbed, it will hang down in a flurry of grey-green leaves and sprays of scarlet flowers.

HALF-HARDY PERENNIALS

Gardeners with warm, sandy soils who live in mild localities may try a number of half-hardy perennial plants whose expectation of life would not be long on heavier ground. Even so, they are a grave risk in cold winters. With plants such as mesembryanthemum, dimorphotheca, gazania and venidio-arctotis, it is wise to keep cuttings in a frame as replacements.

22   *Above:* Mrs Furnival is a rhododendron with rich pink flowers, marked with a darker pink blotch

23   *Below:* Pink Pearl, with beautiful soft pink flowers, is probably the best-known rhododendron

24 *Ulex europaeus plenus*, the double gorse, is a much better form for garden cultivation than the wild plant

25 One of the best of the early-summer-flowering brooms is *Genista cinerea*. It grows very well on poor sandy soil

26 *Below: Cytisus × kewensis* is a dwarf species, particularly suitable for the rock garden or a sunny bank

The succulent 'mezzies', as the mesembryanthemums are called, may be pushed into a sandy soil when every piece seems to root, and in this way I find it easy to build up a good stock. They revel in a hot, dry position and are excellent for poking into the crevices of dry walls.

Because the dimorphothecas have done so well on our calcareous sand on the north coast of Cornwall, we have made a special study of the whole family. *Dimorphotheca ecklonis*, from hotter regions of South Africa than the hardy *D. barberiae*, cannot be regarded as reliably hardy, but this beautiful plant, with narrow petals of icy white backed with purple and a central zone of bright blue, spreads quickly into a large bushy plant before summer is well advanced. *D. hybrida rosea*, which I suspect to be slightly more reliable, has a flower of satiny pink with a contrasting blue central disk, and the daisy flowers are well displayed on long stems. *D. jucunda*, with white petals and a blue central disk, is a trailer and *D. jucunda rosea*, with wine-pink daisies, has a similar prostrate habit.

We shall not be surprised that the gazania, the Treasure Flower of South Africa, is happiest on sharply-drained light soil in full sun. It brings us some of the exotic colours we associate with tropical countries. There are beautiful hybrids to compete with the bright orange *Gazania × splendens*, which can be grown from cuttings late in the summer or from seed early in the year under glass. Plant out the small plants in May when risk of frost is over.

I never tire of recommending the brilliant venidio-arctotis for a sandy border. This cross between *Venidium fastuosum* and *Arctotis grandis* never sets seed, so that the prodigious display of flower is prolonged until frost puts an end to it. A small plant, set out in May, will cover a spread of quite 2 feet and each day the gerbera-like flowers open in the sun to reveal their beautiful colours of pink, lilac, orange and flame. They stand drought well once they are established, but require a little water to get started. Take cuttings early in August in a sandy frame to get them rooted before the winter.

Aster pappei, though strictly perennial, only survives our winters in mild localities. Its clear blue flowers have a contrasting yellow eye and smother a foot-high bushy plant throughout the summer. Small pieces, taken from the main stem in August, soon root in a shady spot in sandy soil. They should be lifted and potted on, to plant out the following May.

Antirrhinum asarina is only perennial on sharply-drained sandy soil where it is dry in winter. This trailing plant of sticky, downy leaves and ivory snapdragons is useful for bone-dry situations and will still be flourishing when other plants have succumbed to drought. It seeds freely.

Odontospermum maritimum (*Asteriscus maritimus*) is delightful during those always difficult months of July and August. The brilliant golden daisies nestle in a mat of downy leaves and, unlike the gazanias, remain open when the sun does not shine. It will come through normal winters, but cuttings in a frame are an insurance against losses.

Othonnopsis cheirifolia is seldom listed in nurserymen's catalogues. They rightly consider it to be on the tender side. It is a splendid plant for droughty places in full sun, the grey, fleshy leaves spreading out to a foot or more to act as storehouses of water so that the rich yellow flowers are unaffected.

Hieracium villosum, one of the Hawkweeds, is a drought-proof plant for very light soils. As the common name suggests, many are weeds, but this plant with woolly leaves and stems that glisten with a silvery down and bright clean yellow flowers on 12-inch stems in June, might be included where few plants will grow, and though not hardy everywhere this does not matter as it will sow itself on a dry sandy soil.

ORNAMENTAL GRASSES FOR SANDY SOILS

I never feel the garden value of ornamental grasses is sufficiently appreciated, though they are much in demand for modern flower arrangements. Their distinctive foliage—blue, yellow or variegated—provides a contrast for shrubs or hardy plants. The blue-leaved Lyme Grass (*Elymus arenarius*) creates a picture of beauty in association with the sand-loving sea hollies. It does not mind how poor and dry is the sandy waste it is required to cover.

Many of our native grasses have variegated forms which are worth a place. Such a one is *Phalaris arundinacea picta*, better known to gardeners as Gardeners' Garters. This is a splendid plant for sandy soil which makes a bright splash of green and white for many months where its invasive tendencies are not a nuisance.

Avena sempervirens is a striking glaucous blue, 2 to 3 feet high, while the little *Festuca ovina glauca* is fine and dainty, a tuft of almost cobalt-blue through which small bulbs may grow. The useful Woodrush, *Luzula maxima*, may be used for suppressing weeds, growing well beneath trees if the situation is not too dry. The imposing miscanthus, its arching ribbon leaves striped with silver in the form *Miscanthus sinensis foliis striatis*, may be used as a specimen plant, 6 to 7 feet high.

Other attractive miscanthuses are *M. sacchariflorus aureus* with yellow-striped foliage; *M. sinensis gracillimus*, very handsome with blue-green leaves that turn in autumn to a pretty amber; and *M. s. zebrinus*, with bars of yellow running cross-wise across the leaves. That grand gardener Miss Jekyll tells us how she used these decorative grasses in her mixed borders of perennial plants with good effect.

Lagurus ovatus, the Hare's Tail Grass, though strictly an annual, obligingly seeds itself from year to year on our hot soil so that one is never without the soft green leaves and delightful furry tails which are ornaments of beauty in the garden or when picked for flower arrangements.

I much admired a plant of *Helictotrichon sempervirens*, growing in the light, gravelly soil of the University Botanic Garden at Oxford. It made a striking blue patch in the November sunshine.

This short list of ornamental grasses for sandy soils is very incomplete, though it gives some small idea of the wealth of interesting hardy plants among them, whose ease of cultivation on poor light soil may recommend them to lovers of unusual plants.

DESCRIPTIVE LIST OF HALF-HARDY PERENNIALS AND GRASSES MENTIONED IN THIS CHAPTER

BOTANICAL NAME	HEIGHT IN FEET	DESCRIPTION	PAGE
Antirrhinum asarina	½	White flowers from June to September	81
Aster pappei	1	Blue flowers from June to August	81
Avena sempervirens (Grass)	2 to 3	Glaucous blue leaves	82
Dimorphotheca ecklonis	2	White daisy flowers with blue-violet central disks all summer	81
hybrida rosea	2	Pink daisies with blue central disk	81
jucunda	Low-growing	White with blue disk from April to October	81
rosea	Low-growing	Wine-pink flowers from June to August	81
Elymus arenarius (Lyme Grass)	4	Blue leaves	82

BOTANICAL NAME	HEIGHT IN FEET	DESCRIPTION	PAGE
Festuca ovina glauca (Grass)	$\frac{3}{4}$	Blue leaves	82
Gazania × *splendens*	$\frac{1}{2}$	Orange flowers in July and August	81
Helictotrichon sempervirens (Grass)	$1\frac{1}{2}$ to 3	Blue-grey leaves	83
Hieracium villosum	1	Yellow flowers from June to August	82
Lagurus ovatus (Hare's Tail Grass)	1	Furry flowers in June and July	83
Luzula maxima (Wood Rush)	1 to 2	Fresh green foliage and browny-green flowers in summer	82
Miscanthus sacchariflorus aureus (Grass)	4 to 5	Golden-striped, green leaves	83
sinensis foliis striatis (Grass)	6 to 7	Central cream stripes on green leaves	82
gracillimus	5 to 6	Blue-green leaves	83
zebrinus	5 to 6	Yellow, transverse stripes on green leaves	83
Odontospermum maritimum	Dwarf	Golden flowers from May to October	82
Othonnopsis cheirifolia	1	Long grey leaves and yellow flowers in June	82
Phalaris arundinacea picta (Gardener's Garters Grass)	3 to 5	Green and white leaves	82
Venidio-arctotis	1 to $1\frac{1}{2}$	Orange, pink, lilac and flame flowers from June till frost.	80, 81

CHAPTER SIX

Rhododendrons on Sand

(CONTRIBUTED BY JOHN STREET)

THERE are two fallacies about soil which cause trouble and disappointment. One is that the heavy soils, the thicker loams and clays, are all alkaline, and the other is that all light soils are acid. There are so many exceptions to these false rules that they are proved wrong. London clay, for example, where it stretches out into parts of Hertfordshire, is acid and is one of the finest soils for growing rhododendrons and azaleas. And many a crop of cuttings of rhododendrons, azaleas, and heathers has been lost because the sand used as a rooting medium was alkaline.

Two extremes of sand are Bagshot Sand, which stretches from Bagshot in Surrey down through Hampshire to the New Forest and across into Dorset, an acid sand where rhododendrons grow well, while Fontainebleau sand, which occurs in France and consists of grains of quartz held together by calcite, a crystallised form of calcium carbonate, is much the same as limestone or chalk. Both the stalactites and stalagmites found in the Cheddar caves are made up of a formation of calcite, and you could not find a more alkaline area than Cheddar.

Acid soil and humus. The first requirement for growing rhododendrons and azaleas is an acid soil. The structure is less important. It might be thought that they prefer a sandy soil, or, for perfection, a sandy-peaty soil, because so many of the commercial nurseries which specialise in rhododendrons are located on Bagshot Sand. But the reason for this is that this particular type of soil is good for root formation; it is also well-drained (which makes it possible to work on it more often than slow-draining soils) and it is easy to work—which saves labour. But for permanent planting, rhododendrons will often grow much better on a heavier soil, although they will need some help in the form of humus on light sands if they are to give their best over a period of years. Unless there is a good supply of natural water, sandy soils can become too dry for rhododendrons. The best solution to this is a liberal dressing of cow manure in the preparation of the ground before planting. Some of the best plants that I ever saw were grown on light sand at Churt in Surrey where the owner of a large garden was able to give each bed a cartload of cow manure. It may seem that this insistance on cow manure is a fragment of traditional folklore handed down by gardeners over the generations. But it does have a particular importance in relation to rhododendrons. It increases the acidity of the soil and it is not too high in a dangerous form of nitrogen which, in other manures, can cause burning of the foliage. It is of great importance on light sandy soils because it retains moisture. I once thought that this was an illusion but I have always found that where cow manure is stacked, the ground nearby is always wet.

Substitutes for cow manure. I am unfortunately only too well aware that this advice—to use cow manure—is a policy of perfection which is becoming more and more difficult to attain. It is easy enough to write about cow manure, but it is much more difficult to buy it. Accordingly, substitutes have to be found. One of the best, both for moisture retention and for suitability to the cultivation of rhododendrons, is spent hops or hop manure. Peat is also one of the rhododendron's staple foods and it will also help to keep the water in the soil, but once it becomes dry it is very difficult to make it wet. When it is used for rhododendrons, either in the preparation of the soil or in mulching, it is important to make sure that it is thoroughly damp before it is dug into the soil or spread over the roots. Garden compost will also help to create a store of moisture, but it must not be prepared with any activator which has any suggestion of alkalinity. Sulphate of ammonia will provide additional nitrogen and, at the same time, give a slightly acid flavour to the mixture.

Preparation of soil. The method of preparing the soil for rhododendrons and azaleas is the same whether it be heavy or light. The ground should be well dug to two spade depths for the whole of the area that is to be planted—a hole taken out in unprepared ground, sufficient only for the size of the roots, is asking for trouble. The additional humus—cow manure, spent hops, hop manure, peat, compost—should be worked into the soil which should then be left for at least a month, preferably longer, before planting. The reason for this is that well dug soil settles and if rhododendrons are planted in it immediately, they are apt to be left with their surface roots above the ground. Unless these are immediately provided with a mulch, they will suffer, and because they are the most active of all the roots the plant will suffer even more.

Root action. The question of root formation becomes vital if there is any risk of drought. It could well be the overriding factor in growing rhododendrons and azaleas on a hot, light soil. It would therefore be the greatest mistake to be dogmatic about grafted plants or plants on their own roots, because a good root action is the best basic answer to lack of moisture.

With the large-flowered rhododendrons of the more popular varieties, Pink Pearl, Cynthia, Britannia, etc., it would be better to obtain grafted plants of top quality. This would mean going to a nursery and selecting the plants in the late summer or early autumn for their appearance at that time of year. By doing this it would be possible to pick out those which have vigorous root systems, and a good union between stock and scion both of which would be apparent in the general health of the plant. Layered plants seldom make the same progress except under perfect conditions.

The same rule does not apply to all azaleas. The older Ghent Hybrids will make good strong plants when they are grafted on to the common yellow azalea, *Rhododendron luteum*, but the Mollis varieties are better on their own roots, and many of the newer Knap Hill varieties do not do well as grafted plants. The variety that is the best (or worst) example of this is the lovely pale yellow Harvest Moon which must be on its own roots.

Preference for sandy soil. In spite of the fact that all the rhododendrons are raised by grafting on the same stock—*R. ponticum*—there is a remarkable difference between the behaviour of different varieties. I would not wish to give the impression that any of them are in any way temperamental about soil, except for the vital requirement that it should be acid, but experience has shown that there are some which will do just that little bit better on a light soil. Two varieties which stand out in my mind above all the rest are Cynthia and Britannia. Cynthia is a rich crimson with dark green foliage set on a somewhat angular branch system. It grows to about 10 or 12 feet in 20 years, from a 15- to 18-inch plant, and if it has a fault it is that the leaves hang down in the winter, especially in periods of cold weather. Britannia is the most popular of all the red rhododendrons. It is as near to scarlet as may be had in a really hardy hybrid. The habit is compact, the leaves are a light yellowish-green and they set off the colour of the flowers, which are of a rich texture and shaped like those of a gloxinia. I have seen both these two varieties growing well on dry banks in Camberley in Surrey, where the Bagshot sand is even lighter than it is at Bagshot itself, only three miles away.

These are two which seemed positively to enjoy a light, hot, sandy soil. The following is a list of some more which also do well in these conditions. However, it should not be thought that there are not a great many of the hardy hybrids which would be perfectly satisfactory on sandy soils. These that I am mentioning seem to me to do better on sandy soils than they do on heavier soils.

PINK

Betty Wormald. A rich pink with deeper spots with a large truss in the same shape as Pink Pearl (described below) and thought by many to be a considerable improvement on that variety. Growing to about the same height, with leaves of similar shape, it is a little susceptible to damage in the foliage if a very active farmyard manure (either pig or horse) is used as a mulch.

Mrs Charles E. Pearson. This variety was bred in Holland as a possible substitute for Pink Pearl in the colder parts of the United States. It opens with a tall bud which is, strictly speaking, a very delicate mauve-pink fading to pure white, with brown spots in the throat. It grows to about the same height as Pink Pearl and has strong, dark green leaves.

Mrs Furnival. One of the sights of the Royal Horticultural Society's garden at Wisley, where it grows vigorously on the very light sand of Battlestone Hill. It forms a rounded bush with good, dark leaves and the flowers are rich pink emphasised with a darker pink blotch. It will grow as tall as Pink Pearl but is more wide-spreading and grows outwards before it grows upwards.

Pink Pearl. This is the best-known rhododendron, with its deep pink buds opening to soft pink flowers. It is a medium grower, reaching 8 to 10 feet in about 15 years, and the habit is upright. There is one interesting fact of cultivation about this variety —not only does it grow well on sandy soil, but it actually prefers a soil with plenty of stones in it.

RED

Joseph Whitworth. This is an old variety with a similar habit to that of Mrs Furnival but with rather larger and darker leaves. The flowers are maroon-red appearing rather late in the season, towards the end of May.

Kluis Sensation. Similar in colour to the well-known crimson-scarlet variety Britannia but rather taller growing and some 10 days later to flower. This variety enjoys a light sand, but because it is late-flowering it prefers a little shade in order that the blossoms should last longer.

Moser's Maroon. The colour is described in the name of this variety. It is a tall grower which has bright red young foliage. It is inclined to become leggy and may need pruning every six or seven years. It is worth growing for the bright colour of the young leaves as well as that of the rich red flowers.
Old Port. This is wine-red, a strong but compact grower which combines very well with Betty Wormald (see page 88) to make an interesting and attractive contrast of colour.

WHITE

Chionoides. A very old hybrid with a name which sounds like that of a species. It is the only compact-growing, late-flowering, white rhododendron. It makes an excellent foil to the common mauve *R. ponticum*, or to the maroon-reds such as Old Port and Joseph Whitworth.
Mrs Tom H. Lowinsky. The rhododendron with the most interesting flower shape. Each individual flower looks like an orchid and the impression is increased by the deep orange flare in the centre.
Mount Everest. A pure white flower with a small red dot in the centre and a faint but definite scent. Like the variety Mrs Furnival, it does very well and makes a fine specimen on Battlestone Hill in the Royal Horticultural Society's garden at Wisley.

YELLOW

Goldsworth Yellow. Another hybrid which does well at Wisley although it is not always quite as vigorous everywhere. The flower buds open pink and turn to yellow. It is a medium grower with a somewhat spreading habit. In many ways it is still the most satisfactory of the many yellows.

I should like to repeat that this is a list of rhododendrons which can be said to prefer a sandy soil. Given adequate preparation and the minimum of aftercare, there are still a great many of the thousand odd hybrids in cultivation which may be grown on any sandy soil, provided it is acid.

DWARF RHODODENDRONS

One of the advantages of gardening on sandy soil is that dwarf plants may be grown, so that they will retain their character. When they are planted in a richer soil they lose their neat habit and become straggly and unattractive. The following is a list of some of the dwarf rhododendrons species which are worth growing, either among the taller kinds, or in rock gardens, or even in mixed borders near the front. They all flower in early May and grow to about a foot in height.

R. calostrotum—large red-purple flowers.

R. fastigiatum—covered in small blue flowers.

R. impeditum—a dense mat of grey-green foliage with a mass of blue flowers.

R. keleticum—flat purple flowers.

R. racemosum—pink flowers on red stems.

R. russatum—a little taller than the rest, deep purple flowers and scented foliage.

AZALEAS

Azaleas are now technically rhododendrons. They have the same botanical characteristics but, in my experience, they need rather different treatment in the garden. Generally, they prefer a slightly heavier soil, but they will grow on sand if it is kept moist and, better still, if they can be so planted as to have their roots in the shade and their tops in the sun. The safest of the deciduous azaleas are the old Ghent varieties and the *occidentale* hybrids. The following is a selection of these, which are known to do well on sand.

Ghent Hybrids (growing to about 4 to 5 feet and flowering in late May).

Bouquet de Flore—pink, striped, yellow blotch.

Coccinea Speciosa—bright orange with prominent stamens.

Daviesii—white, cream blotch, good autumn colour.

Nancy Waterer—bright yellow, good scent.

Narcissiflora—yellow, double, strongly-scented.

Norma—deep salmon.

Raphael de Smet--pale pink, double, scented.

Unique—taller-growing, orange-yellow.

Occidentale Hybrids (grow to about 4 feet, all are beautifully scented, flowering middle to end of May).

Delicatissima—white, with a touch of pink.

Irene Koster—pink.

Superba—pale pink, with a deeper flare in the centre of the flower.

Azalea species. There are a few azalea species which do well on sand. They are named here under their correct designation as rhododendrons.

R. arborescens—very late flowering; white, sweetly-scented flowers and aromatic foliage.

R. luteum (better known as *Azalea pontica*)—a really fine garden plant with well-scented, yellow flowers and brilliant autumn foliage.

R. vaseyi (The Pink Shell Azalea, a common name which describes the
flower)—long pointed leaves in deep bronze.

Mollis Hybrid Azaleas. These are among the best known of all the
azaleas. They are characterised by a predominant colour which is
popularly known as flame, a salmon-pink with a strong touch of
orange. The best plants for a sandy soil—and the least expensive
—are the seedlings. These can either be bought mixed or selected
according to colour, usually in orange, orange-red, red, salmon, pink
and yellow. They come into flower at the beginning of May, before the
leaves appear and grow to a height of about 3 to 4 feet in the course
of 10 to 12 years.

Knap Hill Hybrids. The Knap Hill azaleas are among the best of all
deciduous flowering shrubs. But they have not been grown long enough
in a wide variety of soils to be able to recommend many varieties for
sand. The following are a few which are known to do well. They reach
a height of 5 feet in about 10 or 12 years.

Gog—orange, good grower, good autumn colour.

Golden Eagle—bright orange-red.

Golden Oriole—a fine yellow.

Persil—white, with a yellow blotch.

Redshank—orange-red, with a yellow blotch.

Satan—dark red.

Tunis—orange-red, well scented.

Dwarf Evergreen Azaleas. These make neat bushes about 2 to 3 feet
high (taller in the shade), and are covered with small flowers in early
May. *Azalea amoena* (correctly known as *R. obtusum*)—a neat evergreen growing in layers with flowers of deep magenta except in the
form Coccinea, when they are red.

Betty (Malvatica × *kaempferi*)—large pink flowers, and because
   it is semi-evergreen the old leaves colour well in the autumn.

Esmeralda—shell-pink, very similar to the popular Hinomayo but
   better on a sandy soil.

Hinodegiri—red, with bronze young foliage.

John Cairns (Malvatica × *kaempferi*)—bright red, free-flowering and
   long-lasting.

Palestrina—the best white and the hardiest of all.

AFTERCARE

All help should be given to plants which may at any time have to suffer a period of drought. It becomes even more important to pick the flowers off rhododendrons and azaleas growing on sand than it is with those on a heavier soil.

Watering is the big problem, especially in areas where the supply may be alkaline. This need not be taken too seriously provided the soil is definitely acid, but if it is neutral or alkaline, then only rainwater should be used.

Mulching will do much to overcome drought. Peat and well-rotted cow manure are the best materials. Again it is important to stress the use of cow manure. If pig or horse manure is used on the active young roots near the surface, there is considerable risk of damage to the leaves from too much nitrogen all at once.

Grafted plants must be watched for suckers. Rhododendrons and azaleas do not revert as many people think, but if they are grafted suckers are apt to come from the stock in exactly the same way as they do from fruit trees and roses. They should be chopped off with a spade, making the cut well below ground, as near as possible to the point where it emerges from the main root.

CHAPTER SEVEN

A Garden of Gorse, Broom, and Heather

THOSE of us who admire the flowering of our native moorland plants, gorse, broom, and heather on stretches of sandy heathland, know that these are pointers to shrubs which take naturally to a garden on sand, providing an all-the-year-round display of colour and interest for the minimum of maintenance. Rosarians can turn up their noses, rhododendron addicts will have plenty to say, but the busy gardener or elderly might do worse than give serious thought to such a garden.

A collection of summer- and winter-flowering heathers is an absorbing interest on its own, and the addition of garden varieties of gorse, and a multiplicity of brooms yield an unending succession of bloom throughout the year. The leafy stems of the brooms are too sparse to cast a shade the heathers might resent, and with the taller of the tree heaths they add height to a groundwork of dwarf and spreading heathers.

GORSE

Few will wish to introduce the common wild gorse into the garden, though it provides many a cliff-top garden on pure sand with a fine windbreak against gales and salt, but the freedom with which it seeds can be a nuisance, and we should make use instead of some of the free-flowering garden forms which revel in the poorest soil and will furnish the most arid sandy banks with colour. *Ulex europaeus plenus*, the double gorse, sets no seed and is a great improvement on the wild plant for garden cultivation. In spring a bush is a glorious mass of bright gold along the prickly shoots. It is quite hardy and should be freely planted in gardens on pure sand. The dwarf and compact *U. gallii* very closely resembles it but flowers during August and September, and when grown on poor hungry land in full exposure to sun and wind, is a most satisfactory shrub, retaining the compact form and prostrate habit which it loses on richer soil; in company with other sand-lovers, such as the common ling, it makes a charming display of late colour. Plant gorse in ground which has not previously been disturbed and remember that it prefers a poverty-stricken situation to a rich one.

BROOM

Many a sandy wasteland is made beautiful with the wild broom. Indeed we are apt to regard it a trifle disdainfully for its willingness to grow and seed itself without help from us. It belongs to the pea family, the *Leguminosae*, and there is good reason to believe that leguminous plants form nitrogen in excess of their own needs, sparing some for other plants in their vicinity.

I never plant a broom on our own calcareous sea sand without the thought that the soil is being enriched with much-needed nitrogen. Brooms are well adapted to cope with light, porous sand and heath soils, for they will not tolerate water lying around their roots during the winter months—the merest hint of waterlogging in winter tends to kill them—and because they grow fast and flower at an early age they are much in demand in new gardens on sandy soil. Like most plants of the pea family brooms send down a questing tap root to obtain moisture and thus are able to exist and flower freely on quite unpromising soil conditions.

Their life span can be much increased if regular pruning is carried out from infancy. There seems some confusion about cutting back the brooms. People tell me they are afraid to touch them for fear of spoiling them, so they leave them alone to get gaunt and ugly. Under no circumstances, however, should this essential shaping be omitted after they have flowered, or the bushes will be spoiled and their life shortened. Growth which has just ceased to flower and is setting seed may be cut back to within 6 or 8 inches of the old wood. Secateurs are tedious for cutting the green twiggy shoots and shearing can give too formal a haircut, but a light flick over with a sharp grass hook will do the trick. It ensures a nice bushy appearance for furnishing the winter garden and a wealth of flower the following season. On limy sand, where evergreens are always difficult since ericaceous plants are mostly lime-haters, their green rushy stems are much prized in winter and they were among the few shrubs in our garden which retained their greenery, without turning brown, during the bitter winter of 1962-63.

Planting time for brooms is from October to March, and the addition of some moistened peat will help to establish the young plants in dry, sandy soil. Try to obtain plants grown on their own roots; they grow better and have a longer life than those grafted on to laburnum stocks. Brooms are shocking transplanters. The existence of a tap root means that once planted they must not be disturbed. Sites must be chosen with care, for once committed to the soil they should not be moved again. Named kinds should be bought in pots, never from the open ground, but I like to experiment with my own seed, for many an interesting and beautiful shrub has resulted. Sow the seed in individual peat pots and plant the broom, pot and all, while the plant is still quite small. Cuttings of any special variety may be taken of short, half-ripened lateral growths, with a heel of old wood, in late summer. They root readily in a sandy frame that is not too moist and is exposed to the sun's rays.

Botanically we find the brooms in three genera: *Genista, Cytisus* and *Spartium*, but from the gardener's point of view there is little difference. Perhaps the genista is more tolerant of lime, though we have found all three do well on our sandy soil, which has a *p*H of 7·8 to 8·2. Since all brooms are ardent sun-worshippers they should be freely grown in gardens on sand which cannot be fed continually with humus.

GENISTAS

Genistas vary in habit from quite dwarf, prostrate shrubs to tall, tree-like ones. The tallest is the hardy *Genista aethnensis*, found wild on Mount Etna, from which it gets its name of the Mount Etna Broom. It is a striking shrub, often reaching as much as 15 feet, whose drooping, leafless stems are laden with scented yellow pea-flowers in July, and its tall, pendulous form makes it eminently suitable for the centre of an island bed where its delightfully graceful spread will shelter other dwarfer plants from the scorching rays of the sun without unduly shading them. Give it a firm stake when young.

Genista virgata, the Madeira Broom, grows rapidly to 6 feet or more, and is an altogether more solid bush with yellow flowers produced intermittently from June to October. It is very hardy and will sow itself freely on sandy soil, even establishing itself in roughish grass in semi-shade.

Genista cinerea is one of the best early summer flowerers on poorish sandy soil, growing up to 8 feet or more in a fragrant mass of golden-yellow flowers during June and July. It is one of the most floriferous and beautiful of the taller brooms.

The hummocky Spanish Gorse, *G. hispanica*, is a complete contrast, making a prickly cushion, entirely smothered during May and early June in bright gold. Its habit is round and squat, probably due to the incessant cropping by goats in its native Spain. Plant it on some dry slope, in association with one of the intensely blue rosemaries which flowers at the same time and revels in similar poverty-stricken situations. The Spanish Gorse does not resent being clipped over after it has flowered, and even old bushes may be cut back and will break again.

The popular *G. lydia* is not fussy as to soil and flowers profusely on our light, dry soil. It is reliably hard, at least in the southern half of the country, and fairly cascades down a sunny slope in a shower of yellow in June. *G. tinctoria elongata flore pleno*, a form of our native Dyer's Greenweed with double flowers, gives a wonderful show of orange-yellow at midsummer, and the low-growing *G. pilosa*, though only a few inches high, will clothe a space 2 to 3 feet wide.

CYTISUS

Many a garden on poor, dry sand would make a poorer showing in early spring were it not for the cytisus. There still remains some confusion among newcomers to gardening, between the cytisus (broom) and the cistus (rock rose), and this is understandable since the names have a similarity and each is a lover of hot, sunny situations. The cytisus will tolerate a certain amount of lime and is useful on sand which does not suit the rhododendrons and azaleas. *Cytisus* × *kewensis* is among the earliest to flower, an attractive dwarf, no more than a foot high with a wide spread approaching 4 feet. It is an ideal shrub for clothing a hot, sandy bank where it will flow down in a waterfall of creamy-yellow. At much the same time the very desirable *C.* × *praecox* is a mass of cream flowers, and unlike many of the brooms that tend to get leggy, it is a compact 3-foot bush clothed right to the ground; were it not for the rather heavy smell of its flowers at close quarters it would be difficult to fault it, as it is a decorative furnishing in the garden even when out of flower.

For sheer brilliance of colour, the common broom, *C. scoparius*, is one of the most effective for distant planting and should not be despised for its familiarity on sandy heathland where it seeds itself. Though many of the bicolors are more exciting, the yellow broom makes more show at a distance. Plant from pots without disturbing the roots.

Great as is their beauty in spring, the brooms make dense, shapely bushes of greenery for the rest of the year if trimmed over when they

27 Another of these attractive brooms is *Cytisus × praecox* with cream flowers. It is somewhat taller than *C. × kewensis*

28 *Erica × darleyensis* produces its light purple flowers in the winter, from November to April

29 *Below: Erica mediterranea* follows *E. darleyensis* in flower, and the white-flowered variety W. T. Rackliff provides a display in March and April

30   *Calluna vulgaris flore pleno* is a pretty ling with its double soft pink flowers

31   The tomato-shaped hips of *Rosa rugosa* are a splendid bonus on a plant which already has the merit of flowering throughout the summer

have gone out of flower. Few shrubs need pruning more than the brooms, yet they are often left alone to make gawky, unattractive bushes which are no credit to any garden. Their whippy, rushy stems stand a lot of wind and they should be freely planted in exposed gardens.

Cytisus racemosus, the genista of florists' shops, is nothing like as hardy and only grows happily in the open in sheltered gardens in the south. No one who has seen its glorious masses of golden flower at Tresco in the Isles of Scilly can forget its scent, the sweetest of any broom. It is worth recording that it flowers in January in our own garden on the north coast of Cornwall in the open, surviving 10 degrees of frost, and is so easy from cuttings that it is worth a risk in mild districts in a sunny, sheltered place. The erect and fast-growing hybrid Porlock is more frost-hardy and flowers intermittently from February to May with yellow flower clusters down the long shoots, but has not the phenomenally long season of flower of *C. racemosus*.

SPARTIUM

Not all shrubs which go under the name of broom belong to the genera *Genista* or *Cytisus*. In a class by itself is the Spanish Broom, *Spartium junceum*. This accommodating shrub deserves high marks, since it has a long season of flower from June to October, a scent reminiscent of a bean-field, long spikes of yellow pea-flowers good for picking, and is so easy from seed that a child could grow it. It endures the hottest, driest part of the garden and the poorest of sandy soils. No broom stands severer pruning—you can cut it almost to the ground and it will break again from old wood. If permitted, it sets a prodigious amount of seed, but this is an exhausting process and should not be encouraged.

HEATHS AND HEATHERS

We are all familiar with the heath (*Erica*) and heather (*Calluna*) of sandy heathland though we need not concern ourselves here with the botanical difference which exists between the two. Garden forms of heathers become increasingly popular as gardeners appreciate the true value of these easy, long-suffering plants. They require the minimum of care once they are planted, since they are splendid weed-smotherers when they have closed their ranks, and even the trimming over that is often recommended may be neglected when they are grown in full exposure to sun and wind. They are as frost-hardy as any plant I know and the winter-flowerers appear from beneath a covering of snow .

ice as vigorous and beautiful as ever. Perhaps their most endearing quality is that, when planted in variety, there is scarcely a dull moment in the heather garden, since one or other will be in bloom. For the remainder of the year they have delightful green foliage, while many newer kinds are reddish-bronze or gold.

Gardens on acid sand are fortunate that these provide ideal conditions for growing heathers. On notorious Bagshot heathland there is a famous heather garden where many exciting novelties may be seen, and one has only to visit the heather garden at Windsor Great Park, where roughly seven acres are given up to many varieties, to realise how happy is a heather garden on sandy, peaty soil, for this was once a gravel pit and the soil is sand and gravel. Here are found most of the summer bloomers as well as those which flower in winter and many daboecias (including *Daboecia cantabrica*, the Irish Heath, and its varieties), sometimes shades of purple, red, pink, and sometimes white, with large, egg-shaped bells.

Unfortunately there is always one reservation when recommending heathers: all but one of the summer flowerers dislike lime in the sand, the exception being the Corsican Heath, *Erica terminalis* (*E. stricta*). However, since acid sand is far more prevalent in this country, most of us do not have to concern ourselves with this restriction. Heathers, I feel, should not be relegated to some obscure corner of a garden, rather should they be a feature of the garden on sand which is exposed to sun and wind. That there are heathers in varying heights and spreads does much to relieve the sameness of the heather garden, for there are tiny plantlets as well as tall tree-like heaths.

The heather season begins on any soil, acid or alkaline, with the winter flowerers, the bush *Erica* × *darleyensis* with sprays of light purple flowers, *E.* × *d.* Arthur Johnson with longer and brighter sprays, and the indispensable 'carneas'. Their soft pinks, carmines and bright reds defy the coldest winds or the severest frosts, and pools of colour emerge as the snow melts to reveal their brave and lovely faces. Busy gardeners are turning more and more to these labour-saving plants. I am often accused of being biased in favour of the Springwoods, and no heather garden can afford to be without the low, spreading, *Erica* Springwood, white, and Springwood Pink. They are grand as irregular edgings to a path or massed along a drive, for they look their best in bold drifts of one variety.

Scarcely have the winter flowerers passed their best (three to four months of bloom is not unusual) than the dense and bushy *E. mediter-*

ranea superba is showing pink. This makes a most lovely 4-foot bush of deep green foliage, each branch being wreathed with a profusion of rose-pink flowers from March to May. *E. m.* W. T. Rackliff is a hardy, compact, white form, not more than 2 feet in height, in flower during March and April.

From now until *Erica terminalis* flowers in August, a tall heath often 5 to 6 feet high which makes a splendid hedge, we gardeners on calcareous sandy soil must say good-bye to heathers. Gardeners on acid sand will, however, have the whole range of summer-flowerers to choose from. *E. cinerea* likes an acid moorland sand laced with peat and an open exposure, for the bell-heathers are sun-lovers. Two of the best are C. D. Eason in glowing red and the pink Apple Blossom. Forms of the Dorset heath follow on and if I had to choose it would be the tall and robust *E. ciliaris* Stoborough. Though *E. ciliaris* is more at home in the south, *E. tetralix* is happy in the north on wet sandy marshes. Its habit is short and close, and of these Con. Underwood is one of the best, a deep glowing crimson against silvery foliage. Through August and September the beautiful Cornish heaths are in flower, well represented by the white *E. vagans* Lyonesse and the salmon-pink *E.v.* St Keverne, and these carry through into the autumn, bearing the lings company. There are wonderful garden forms of our common ling, *Calluna vulgaris*. The vigorous H. E. Beale has foot-long sprays of double pink flowers, while pink Rosalind has an extra bonus of golden foliage. C. W. Nix has crimson flowers and the white *serlei* prolongs the season. *C. v. flore pleno* is a double ling with pretty flowers of soft pink.

Lovers of pretty foliage will plant the winter-flowering *E. carnea vivellii* whose dark bronze leaves turn back to green in summer, *E. cinerea* Golden Hue, with light gold foliage which turns a reddish shade in winter, *Calluna vulgaris* Gold Haze, with bright golden leaves to offset its pale lilac flowers, and *C. v.* Robert Chapman with foliage of exceptional beauty, turning through gold, flame, bronze and orange throughout the year. It is in short supply but will become increasingly popular as it becomes better known. The grouping of these foliage heathers needs special care; plant them facing south or south-west, never towards the north or north-east. The sunnier the aspect the brighter will be the foliage—it is the sun which keeps them brightly coloured.

Heathers, like rhododendrons, are surface-rooting, so plant shallowly but firmly. The plant should appear to be sitting comfortably on the ground, and use plenty of peat. Planting can be carried out from the beginning of October until the beginning of April, the sooner the better so that plants have a chance to get established before a dry spring sets in.

Tree heaths appear to have escaped the attention which the dwarf heathers have received, gardeners supposing them to be too risky in our climate. *E. arborea* is probably too tender except for mild districts, though *E. a. alpina* came through the icy blasts of our coldest winter without appreciable damage. It has ashen-white flowers and foliage of vivid green, on a bush 5 to 7 feet tall. *E. lusitanica*, with lovely wreaths of white flowers touched with pink in mid-winter, seeds itself on sandy soil free of lime, and though the woody stems are sometimes split by frost, the roots invariably survive and the shrub breaks anew. The branches are a lovely decoration for the house. Although the Spanish heath, *E. australis*, is one of the loveliest of tree-heaths, with large bells of rosy-red, it is only hardy in warm, sheltered gardens. *E. a.* Mr Robert is a white-flowered variety that is similar in habit, but the foliage is a paler green and it is reputed to be tougher than the type. The best way to plant these tall heaths is in broad masses so that one shrub protects another.

Such a garden, of gorse, broom and heather would satisfy even the collector's appetite, giving a prolonged season of flower for the minimum of effort, since plants never do as well as on a soil they love.

DESCRIPTIVE LIST OF PLANTS MENTIONED IN THIS CHAPTER

BOTANICAL NAME	HEIGHT IN FEET	DESCRIPTION	PAGE
Calluna vulgaris C. W. Nix	2	Crimson flowers in August and September	99
flore pleno	1½	Double, pink flowers from August to to October	99
Gold Haze	2	Pure white flowers from August to September; golden foliage	99
H. E. Beale	2	Long sprays of double pink flowers from August to October	99
Robert Chapman	1½	Foliage of outstanding beauty and soft purple flowers from August to October	99
Rosalind (Underwood's Variety)	1½	Pink flowers from August to September. Golden foliage	99

BOTANICAL NAME	HEIGHT IN FEET	DESCRIPTION	PAGE
Calluna vulgaris serlei	2	White flowers in September and October	99
Cytisus × *kewensis*	1	Creamy-yellow flowers in May	96
praecox	3	Cream flowers in May	96
racemosus	6	Golden flowers most of the year	97
Porlock	6	Yellow flowers from February to May	
scoparius	5 to 6	Yellow flowers in May	96
Erica arborea	5	White flowers in March and April	100
alpina	5	Ashen-white flowers in March and April	100
australis	6	Rosy-red flowers from March to June	100
Mr Robert	6	White flowers from April to June	100
carnea	Dwarf		
Springwood		White flowers from February to April	98
Springwood Pink	Dwarf	Pink flowers from February to April	98
vivellii	Dwarf	Red flowers from February to March; bronze foliage	99
ciliaris	2		
Stoborough		White flowers from July to October	99
cinerea	1 to 2	Rosy-purple flowers from June to September	99
Apple Blossom	1	White flowers, pink-flushed, from June to August	99
C. D. Eason	$\frac{3}{4}$	Red flowers in June and July	99
Golden Hue	1	Pink flowers in June and July; golden foliage	99
darleyensis	$1\frac{1}{2}$	Rose-purple flowers from November to May	98
Arthur Johnson	$1\frac{1}{2}$	Dark pink flowers from December to April	98
lusitanica	5 to 10	White flowers from December to April	100
mediterranea superba	4 to 6	Rose-pink flowers from March to May	98, 99
W. T. Rackliff	2	White flowers in March and April	99

BOTANICAL NAME	HEIGHT IN FEET	DESCRIPTION	PAGE
Erica terminalis	5 to 6	Rose-pink flowers from July to September	99
tetralix	$\frac{3}{4}$		
Con. Underwood		Crimson flowers from June to October	99
vagans	$1\frac{1}{2}$	Salmon-pink flowers from August to September	99
St Keverne			
Lyonesse	$1\frac{1}{2}$	White flowers from August to October	99
Genista	15		
aethnensis		Yellow flowers in July	95
cinerea	8	Golden-yellow flowers in June and July	95
Genista hispanica	2	Golden flowers, May and June	95
lydia	2	Yellow flowers in June	96
pilosa	$\frac{1}{2}$	Bright yellow flowers in May and June	96
tinctoria elongata flore pleno	Dwarf	Double orange-yellow flowers in June	96
virgata	6	Yellow flowers from June to October	95
Spartium junceum	6 to 8	Yellow flowers from May to November	97

CHAPTER EIGHT

Good Roses for Bad Sandy Soil

THAT 'Roses prefer a clay soil', or that 'Roses do best on clay', are such commonplaces we have almost begun to believe them; but what roses really want is good drainage and a stiffish loam with plenty of humus to retain moisture. The garden on sand can provide the former but sadly lacks the latter, unless it has been incorporated as rotted manure, garden compost or peat laced with a general fertiliser, all of which help to build up the fertility as well as give the sand body.

For some reason which I find difficult to understand, many gardeners insist on making roses the main feature of their garden, regardless of the soil they have to offer. On excessively sandy soil they struggle to grow them against insuperable odds, and I can think of few things more miserable than the pitiable objects we see striving against impossible difficulties for survival. It is true that perseverance, some skill and a deep pocket work wonders, though I am convinced that it is wiser for the ordinary gardener to choose plants that suit his soil rather than try to make the soil suit the plants, and this is particularly true of roses. If we understood the subtle affinity which links a plant to its chosen soil we should be more than half-way to mastering most of the difficulties of cultivation.

Improvement of soil. Although the ordinary bedding roses may be adaptable and long-suffering, there is plenty of evidence to show that they are rarely a success on thin, hungry soils such as sand. Where the sand is acid the chance of success is greater, for roses prefer a soil either neutral or on the acid side, but any sand is infertile, and unless nursed is unsuited to roses. If you want to grow good roses on calcareous sand you must give them exhibition treatment; double-trench the rose bed, sow a crop of mustard on it and sprinkle it with sulphate of ammonia to hasten decay—sulphate of ammonia is excellent for alkaline soil as it increases acidity—dig the crop in, and incorporate some bulky manure, or use plenty of peat-moss, which is splendid for root formation.

However, this is still not enough, as roses in such circumstances need considerable aftercare to provide blooms as good as those produced on bushes growing in naturally good rose soil. They are gross feeders, as they take away from the soil and do not put anything back into its place; on sand it is almost impossible to overfeed them. Aftercare consists of exhibition feeding—an application of fertiliser at Christmas when people usually forget about roses—then the basic treatment of a general fertiliser when they are pruned, followed by a mulch of manure, peat-moss or one of the excellent processed manures, available in bags from any good garden store. The mulch can be applied when the soil has really warmed up, from mid-April to mid-May, according to district, and it should be put on when the ground is wet, as the conservation of moisture is its chief purpose.

Though there is no specific evidence to show that a mulch prevents black spot, it may help to prevent the fungus spores from rising, though there is no doubt that by preventing dryness at the roots it tends to reduce the likelihood of mildew. The mulch should be followed by another application of general rose fertiliser in June. In addition, on calcareous sandy soil, rose leaves show definite signs of chlorosis, and the plants should be annually dosed with Sequestrene in spring.

Whether all this trouble and expense yields sufficient dividends or whether it would be wiser to grow roses other than hybrid teas, which succeed on hot, dry soils, is a matter only the gardener himself can decide. Such roses are as beautiful in their own way as the hybrid teas and reflect credit on their grower at comparatively little cost in time and money.

Mr Graham Thomas, who gardens on sand and knows more about roses than most of us will ever learn, tells me he would not expect to grow prize hybrid tea blooms on sand, though the strongest floribundas and shrub roses thrive. This I have found to be so true that we have given up bedding roses on our very alkaline sand with the exception of a few robust floribundas, and are devoting ourselves to the very beautiful and easy shrub roses. These are so successful that I recommend other gardeners with excessively sandy soil to try them.

SHRUB ROSES

Many shrub roses, old and new, succeed on poor, thin soils lacking in humus—this is not to say that they do not respond gratefully to any help you give, but they do quite well without. They ask the barest minimum of upkeep and lend themselves to the kind of informal gardening which many of us are adopting at the present time. Since our garden is unsuited to the ordinary bedding rose, both in soil and environment, I cannot do better than recommend those shrub roses we have found to succeed on our difficult sandy soil.

Thicket-forming Scots roses or Burnets, forms of *Rosa spinosissima*, thrive on hungry sand; there are large areas of this type of rose to be seen on the sand dunes near Newcastle, Co. Down. They also flourish on indifferent soil on the moors of Scotland, on Beachy Head, and in Wales to give only a few instances. Their compact, thorny thickets are dotted all over in May and June with single or double roses of lemon-white. The old, double, yellow Scots rose is always reliable, never failing to flower abundantly, and there are a number of small thicket roses of the same kind. Many sucker freely and are unsuitable for small gardens, though where they can be given a sandy waste to cover, the small burnet foliage and the little roses in white, cream, pink and crimson are very attractive and ask nothing of the grower.

Modern hybrids. In recent years the German breeder, Wilhelm Kordes, has been remarkably successful in crossing various species with some outstanding hybrid roses, so that we now have beautiful modern shrub roses which do well on a meagre soil, particularly when mulched in spring to prevent the soil around their roots from drying out during the summer. They vary in height from 3 to 6 feet but do not grow as large as on richer soil. They are very hardy and do not appear to suffer frost damage to the same extent as do the hybrid teas. None of ours was damaged during the severe winter of 1962-63. They thrive on light sandy soil which would be totally unsuited to the formal bedding roses, and though they are not immune from the diseases that we associate with roses, their vigour enables them to withstand attacks better than roses of more moderate growth. Pruning, which takes up a good deal of the busy gardener's time, is nothing like so arduous; only dead wood needs removing and branches which have suffered frost damage should be shortened.

One of the most beautiful of the Scots hybrids is Frühlingsmorgen; its single, rich, pink flowers have maroon stamens in a yellow centre. Frühlingsgold, also with Scotch ancestry, makes a larger bush of arching branches, studded with a wealth of golden-yellow, semi-double flowers during May and June. Frühlingsduft, whose fully double flowers are cream flushed with pink, and Frühlingsanfang, with large, single yellow flowers, also suit the same conditions.

Two modern shrub roses excellent on sandy soils are Elmshorn and Bonn. Elmshorn is particularly free, with large clusters of small roses in a shade between carmine-red and pink, while Bonn is a most vigorous rose whose sweetly-scented flowers are an unusual and telling orange-scarlet. Two old roses fit to bear them company are Stanwell Perpetual, with dainty burnet foliage, and a virtually non-stop succession of sweetly-scented semi-double blush-pink flowers from June to October, seldom exceeding 4 feet on sandy soil, and *R. villosa duplex*, the old 'Wolley-Dod's Rose'. The foliage is soft grey-green and the large semi-double flowers are rich rosy-pink, followed by large, hairy, red, rounded fruits. It makes a sturdy bush and is in flower during May and June.

Rugosa hybrids. *Rosa rugosa* has been pretty well submerged under the flood of hybrids it has yielded, but it still remains a valuable plant for extremely poverty-stricken conditions of pure sand. It appears not to care how meagre and barren is its situation; the scented single flowers of purplish-rose keep coming throughout the summer, and one of the joys of autumn are its ornate red hips. The colourful autumn fruits of the rugosas are outstandingly beautiful and since they are inclined to drop off where the soil is extremely poor, they may be picked with impunity and are much in demand for flower arrangements when flowers are scarce. It is not surprising that there is a white variety, *R. rugosa alba*. It has more glossy and more deeply-veined and attractive foliage than the type, and the single snow-white roses are large with yellow stamens, followed by large orange-red fruits. The habit of the rugosas is an attractive one, bushy and clothed right to the ground. They are superb weed-smotherers, vigorous and healthy and undemanding of attention.

For excessively sandy gardens close to the shore it would be hard to find a better rose than *R. rugosa scabrosa*, with most beautiful foliage which stands direct sea-wind, and clusters of extra large flowers of bright rugosa pink with typical creamy stamens and attractive large orange hips for which it was given an Award of Merit recently by the Royal Horticultural Society as a fruiting plant. One of the loveliest of the hybrids is our old favourite, Conrad F. Meyer, a vigorous pillar rose which flowers profusely in May and again in September and October, with enormous double roses of silvery-pink that are very sweetly scented. These are carried on long firm branches so that it should have a post or pillar to support it, or a position at the back of the shrub border where it can tower over dwarfer companions. All this loveliness is guarded by the most vicious prickles, so that I have known it make a fine impenetrable hedge, deterring the most persistent invader.

No garden on sand should be without the cold, white beauty of Blanc Double de Coubert. This is a most beautiful rugosa, whose heavily-veined green foliage is so tolerant of storm and tempest, and is one of the finest we have for exposed gardens on the coast. It forms a magnificent, large, rounded bush with clusters of dead white, semi-double flowers. Its low spreading habit makes it entirely suitable for clothing a difficult sandy bank.

I do not know a rose with sweeter scent than the deep burgundy flower of Roseraie de l'Haÿ. This is a splendid wind-tolerant shrub for poor, dry soils. Its rich peony-like flowers are borne in great profusion over many months. The lovely Frau Dagmar Hastrup is a lovely shrub rose for any garden; it has masses of clear pink flowers followed by rich crimson, tomato-shaped fruits. Its compact habit makes it a good rose for hedging purposes, since it still produces flowers and hips in great profusion even when lightly pruned. The fragrant Snow Dwarf, often listed under the unattractive name of Schneezwerg, is a neat bush of moderate size, smothered all summer and autumn with semi-double white roses, suggestive of the Japanese anemone, whose second crop of bloom often coincides with glossy orange hips. Agnes is a creamy-buff with a wonderful scent and Mrs Anthony Waterer has very heavily scented, semi-double, magenta-crimson flowers, while the semi-double Sarah Van Fleet is fresh pink.

In the rugosa group the Grootendorsts are a little out of the ordinary; they have very small, fully double flowers with frilled edges, which bear a distinct resemblance to a carnation. Though individually the double flowers are not large, their colours, bright red in one and clear soft pink in the other are attractive and cheerful. For small gardens the red F. J. Grootendorst and Pink Grootendorst are among the most useful and decorative members of the rugosa hybrids for poor sandy soils. They make rounded bushes of about 4 feet, amply furnished with flowering shoots annually refurbished from the base. The broad green foliage is remarkably disease-free and the flowers, borne in copious clusters, keep coming from mid-summer till December on our difficult limy soil.

To sum up: the rugosa hybrids are indestructibly hardy, recurrent to perpetual-flowering, from late spring to late autumn, many with scented blooms and an abundance of bright hips to follow, and all with handsome rugose foliage which turns brilliant yellow in the autumn.

The China roses. The introduction of the China rose (*R. chinensis*) has had a tremendous influence on the modern rose because of its delightful perpetual-flowering habit, and on the same meagre soil which supports the *rugosa* and the *spinosissima* roses, the old-fashioned China or Monthly Rose is content with little but gives much. Indifferent to the poor, dry root-run which is all sand provides, they are true perennials in soils quite unsuited to the decoratives, succeeding in quite light, hot ground, and this without fussing about mulches and fertilisers. Their foliage is evergreen and singularly immune to pests and diseases, and they require little pruning.

They are a convenient size for limited areas and in addition to their persistent blooming their colours, unlike that of so many roses, increase in depth by exposure to sun.

The pale silver-pink, semi-double flowers of that grand old favourite, the Old Blush Rose, change to a light rose-crimson. Hermosa is a soft milky pink and rarely out of flower. Then there is Fellemberg, a most vigorous bush of bright red-crimson roses from early summer into winter.

All these are hardy on a light, meagre soil.

FLORIBUNDA-TYPE ROSES

There are many who regard the floribundas as sufficiently tough for poor sandy soils, but such has not been my experience on very alkaline sand. Only the most robust should be considered. On the other hand, the hybrid musks roses put up with the most niggardly soils and are full of flower early and late. They are medium-sized shrubs with large clusters of roses in a wide variety of colours. Penelope, Felicia, Wilhelm, and Will Scarlet are all beautiful, and Thisbe is a soft yellow, of great value in the garden as it blends happily with all the other colours. All the hybrid musks have a sweet scent which is 'free on the air'. If flower clusters are cut back to a strong bud as soon as they are over, a second and even a third crop of flower will follow.

CLIMBING AND RAMBLER ROSES

Only the most vigorous of climbing and rambler roses do well on very sandy soils. The places where they grow are often excessively dry and though the ground may have been dug deeply and prepared with plenty of rich organic material at the beginning, after a year or two the plants will have used up all the nourishment there is and begin to deteriorate. Copious watering and feeding with liquid fertilisers are essential to keep them in reasonable health.

If you have a new house with a wall to cover, plant the thornless rose, the very sweetly scented Zéphirine Drouhin. It flowers almost non-stop and is a very thrifty doer in bad conditions; it does not grow too tall to be a nuisance, 9 feet is about its limit. Albertine is a splendid rambler, with coppery-pink, large, scented, double blooms, opening in mid-June, but its season of flower is regrettably short. On the other hand, soft coral-pink Thelma has a very long season of bloom, its immense clusters hanging for quite two months at mid-summer. It is not too demanding, and a young plant is already doing its best to conceal our too obvious oil-tank. Chaplin's Pink Climber has a longer reach and has nice glossy foliage and large trusses of bright carmine flowers with a white eye, with some resemblance to its parent, American Pillar, and inheriting its vigour without its brashness.

ROSE HEDGES ON SAND

Roses make enchanting hedges though only a few are suited to very sandy soils. One cannot go wrong, however, with the *rugosa* types and hybrid musk roses. The *rugosa* roses make lovely hedges, which have the advantage of being evergreen in most climates, and cannot be surpassed for extremely exposed situations close to the sea where they receive every salt wind which blows. Their heavily-scented flowers are borne through the summer and autumn; the veined, rugose foliage seems impervious to disease, and the single varieties carry handsome red hips in autumn. Blanc Double de Coubert, Roseraie de L'Haÿ, *R. rugosa alba* and Frau Dagmar Hastrup are all good hedgers and should be planted 3 feet apart.

Hybrid musk roses make hedges 4 to 5 feet tall, flowering for two months at mid-summer and again in the autumn. Felicia, Moonlight, Will Scarlet and Prosperity are excellent and should be lightly pruned after flowering, with more drastic treatment in early spring to keep them dense and free-flowering. Plant 3 feet apart.

Dig the site well, remembering that these hedges will continue to be in one place for years, giving them as much help in the way of organic material and peat laced with bonemeal as possible, to get the roots moving. They will welcome an annual mulch each spring but will do quite well without.

DESCRIPTIVE LIST OF PLANTS MENTIONED IN THIS CHAPTER

BOTANICAL NAME	HEIGHT IN FEET	DESCRIPTION	PAGE
Rosa Agnes	6	Creamy-buff, double flowers from July to September	107
Albertine	25	Copper-pink, double flowers in June	109
Blanc Double de Coubert	4	White, semi-double flowers in May	107, 109
Bonn	4	Orange-scarlet, semi-double flowers May to August	106
Chaplin's Pink Climber	10 to 15	Bright pink, semi-double flowers from June to October	109
chinensis (The Monthly Rose)	4	Crimson-pink flowers from June to September	107
Conrad F. Meyer	6 to 8	Silvery-pink, double flowers in May, and again in September-October	106
Elmshorn	3	Double, carmine-red flowers in large trusses June to November	105, 106
Felicia	6	Salmon-pink, semi-double flowers from June to December	108, 109
Fellemberg	6	Double crimson flowers in clusters from June to December	108
Frau Dagmar Hastrup	4	Pale pink, single flowers from June onwards and good autumn fruits	107, 109
Frühlingsanfang	3 to 4	Single, creamy-yellow flowers in May and June	105
Frühlingsduft	3 to 4	Double, cream, flushed-pink flowers in May and June	105
Frühlingsgold	4	Golden-yellow, semi-double flowers in May and June	105
Frühlingsmorgen	3 to 4	Rich pink, single flowers in May and June	105
Grootendorst roses	4	Small pink or red flowers with frilled edges from June to December	107
Hermosa	2½	Milky pink double flowers from June to December	108
Moonlight	6	White, tinted with lemon from June to October	109

BOTANICAL NAME	HEIGHT IN FEET	DESCRIPTION	PAGE
Mrs Anthony Waterer	6	Magenta-crimson, semi-double flowers from June to October	107
Old Blush	3 to 4	Silvery-pink flowers from June to December	108
Penelope	5	Shell pink, semi-double flowers in clusters from June to September	108
Prosperity	6	White flowers from June to October	109
Roseraie de l' Haÿ	6	Crimson-purple, double flowers from May to November	107, 109
rugosa	5 to 8	Single, purplish-rose flowers in June and July and red fruits	106
alba	5 to 6	White, single flowers in May and June	106, 109
scabrosa	4	Single, rosy-carmine flowers in May and June	106
Sarah van Fleet	8	Pink, semi-double flowers in clusters from June onwards	107
Snow Dwarf (Schneezwerg)	4	White, semi-double flowers June to September	107
spinosissima (Burnet or Scots Rose)	1 to 3	Lemon-white flowers in May and June	105
Stanwell Perpetual	3	Pink, semi-double flowers from June to October	106
Thelma	6	Coral-pink, double flowers in clusters in June and July	109
Thisbe	4 to 5	Buff-yellow, semi-double flowers from June to December	108
villosa duplex	6 to 8	Semi-double, rosy-pink flowers in May and June followed by red fruits	106
Wilhelm	3	Crimson, semi-double flowers from June to December	108
Will Scarlet	5	Crimson-scarlet, semi-double flowers in large clusters from June to December	108, 109
Zéphirine Drouhin	8	Carmine-pink, semi-double flowers from July onwards. Thornless.	109

CHAPTER NINE

Trees for Sandy Gardens

'He that planteth a tree is a servant of God,
He provideth a kindness for many generations,
And faces he hath not seen shall bless him.'

MANY trees are at home on poor, sandy soils, particularly the useful conifers, though all are not suitable for the modern garden, and future generations will not bless us should we plant one which outgrows its situation.

The Silver Birch (*Betula pendula*) seeds itself freely on light, well-drained soil, and is one of the commonest native trees, but 'The Lady of the Woods', though one of the most beautiful and graceful of trees, is better left to the woods and hillsides rather than absorbing nourishment meant for other plants in the small garden. Where there is space, however, for its rapid growth, its dappled shade allows the underplanting of carpeters such as heathers and bulbs, and I would not hesitate to recommend it for a sandy soil.

The magnificent Beech is good on alkaline sandy soil and the Sweet Chestnut thrives on a hot, dry soil that is free of lime, though each is too large for the purpose we have in mind. Probably no pine has been more widely planted on pure sand than the Corsican, (*Pinus nigra calabrica*) which, with *P. contorta*, has been largely used for stabilising the Culbin Sands in Moray, Scotland and elsewhere. *P. pinaster*, the Maritime Pine, is very tough on exposed coastal sands, while the Scots Pine, *P. sylvestris*, is one of the hardiest and least exacting of trees, making do with even shallower soil than the larch or spruce, and often found on sandy heathlands. *Populus albus*, the White Poplar, comes second to none of the hardwoods for planting on sandy wastes, affording welcome shelter for less hardy kinds, and the Common Elder, *Sambucus nigra*, is valuable for planting on sand and shingle where it thrives in a remarkable manner.

We shall confine ourselves here, however, to some trees more suited to the garden of today.

32 *Rosa rugosa scabrosa* grows extremely well on sandy soils, and stands up to sea-winds without appreciable damage. The flowers are pink with creamy stamens and it bears attractive large orange hips

33 Conrad F. Meyer is a *rugosa* hybrid pillar rose with silvery-pink flowers

34 *Above:* The pink-flowered Frau Dagmar Hastrup is a good rose for hedging purposes. It bears rich crimson fruits

35 *Above:* The amelanchiers are a genus of small flowering trees or shrubs, very suitable for the smaller garden. The leaves turn red in the autumn

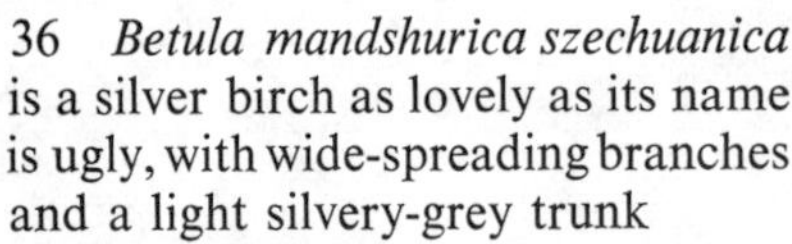

36 *Betula mandshurica szechuanica* is a silver birch as lovely as its name is ugly, with wide-spreading branches and a light silvery-grey trunk

Acacia, False (See ROBINIA).

Amelanchier (Snowy Mespilus). This is a genus of small hardy trees which thrives in any soil. If you have room for only one, plant *A. asiatica*, an elegant small tree of about 15 or 20 feet, bearing an almost incredible profusion of pure white flowers in May and with rich red autumn colour.

Arbutus unedo. Though the Strawberry Tree is indigenous on the peaty soil around the Lakes of Killarney in Ireland, this is a rare example of one of the ericaceous family which grows well where there is lime, and it is doing extremely well on our calcareous sand. It is sufficiently hardy for the southern half of England, and its red stems, glossy wind-tolerant leaves and white pitcher flowers, borne simultaneously with the rough orange-red fruits in late autumn, are a never-failing source of interest. The variety *rubra* lacks the stature of the type but none of its beauty. It is not a very tall tree, reaching about 15 feet.

Betula pendula youngii (Weeping Birch). This, with the weeping pear, *Pyrus salicifolia pendula,* is perhaps the best of all weeping trees for modern gardens. It grows slowly to about 30 feet and its branches fan out and down to hang vertically from the beautiful, pale grey trunk. The wide-spreading branches of *Betula mandshurica szechuanica* from Western China lend themselves to the planting of many bulbs which appreciate light shade.

Birch, Weeping (See BETULA).

Blue Cedar (See CEDRUS).

Blue Gum (See EUCALYPTUS).

Caragana. The semi-pendulous *Caragana arborescens lorbergii,* no more than 10 feet high on hungry sand, is a graceful tree with remarkable light green, fennel-like foliage and a quantity of golden pea flowers in May. Give it shelter from wind.

Cedar (See CEDRUS).

Cedrus atlantica glauca. Though the beautiful Blue Cedar cannot be recommended for a wet clay, it does extremely well on a sharply-drained, sandy soil. It is an excellent specimen tree for a lawn, where there is sufficient room for its wide spread.

Cercis siliquastrum. The Judas Tree is a real lime-lover and we should not hesitate to try it on alkaline sandy soil, giving it generous treatment at the start in the way of humus. It will grow to about 25 feet. The pinky-purple pea flowers wreathe the bare branches in May before the large, heart-shaped leaves unfold. It loves the sun and is happier in the south than in the north.

Chilean Tree Myrtle (See MYRTUS).

Cotoneaster. Cotoneasters are indestructibly hardy and easy on sandy soil, acid or alkaline. A few make delightful small trees laden with bright berries. *Cotoneaster frigidus* is a fast-growing vigorous one, 15 feet or more high according to soil, with a spreading habit and loaded with clusters of crimson berries in autumn and winter. *C.* × *hybridus pendulus* is the ideal small weeping tree, if trained on a 6-foot stem, with long cascading branches carrying clusters of white hawthorn flowers followed by brilliant red fruits. *C.* Cornubia is a very vigorous small tree, up to 15 feet tall on sand, whose berries, carried in large clusters, weigh down the branches.

Crataegus (May). Among the most ornamental of the Mays are the double rose-red *Crataegus oxyacantha coccinea plena* and *C. o. maskei* with double, rose-pink flowers. Poor soil produces the most abundant show of flower.

Cupressus macrocarpa lutea. Though scarce and costly, the golden cupressus is the most salt-tolerant of the cypresses for coastal seashore gardens. It grows slowly to 20 feet. The first years are the most difficult and young trees require plentiful supplies of water.

Cupressocyparis leylandii. This is a first-rate tree for the sandy garden, not demanding much from the soil. It is reputed to be the only cypress able to withstand the violent gales in the northern isles. It may be too large for the smaller garden, but is a valuable and attractive hardy tree where there is room for it.

Davidia involucrata. This is a spreading tree, of up to 20 feet, with deeply veined leaves. It is conspicuous in May when draped with pairs of white bracts which have given it the name of the Pocket Handkerchief Tree.

Eucalyptus. I never feel we make the best use of the unique and beautiful eucalypts with their brilliantly coloured, newly-opened leaves, their distinctive stems, and miniature fountains of creamy-white flowers. They are peculiarly suited to modern gardens as they benefit from severe pruning in early spring, which retains the glaucous juvenile foliage and limits the bush or small tree to a reasonable size. Choose a site sheltered from wind and plant young from pots—small plants not larger than a foot and a half are amply large enough—and stake early. Since the fastest growth is in August and September plants are better in the ground before the end of July. Eucalypts are deep-rooting and a sandy soil should be deeply disturbed and made retentive of moisture by the addition of good, rotted vegetable matter. No animal manures should be given them, though some bonemeal is good on sandy soil.

In our own sandy garden we planted the hardy *Eucalyptus niphophila* on a badly prepared, intensely dry southerly slope, where it suffered from drought. Moved to another part of the garden and given a plentiful supply of humus, deeply dug soil and kept well watered, the transformation was quite remarkable. Each day the plant appeared to grow.

Good eucalypts for our climate are: *E. niphophila, E. perriniana, E. gunnii, E. urnigera* and *E. pauciflora.* Though I have been unable to find the first-mentioned either in Bean's *Trees and Shrubs Hardy in the British Isles* or in the Royal Horticultural Society's *Dictionary of Gardening*, this euclayptus, occurring naturally in the Snowy Mountains in southern New South Wales, at an altitude of over 6,000 feet, is certainly one of the hardiest in this country and one of the most wind tolerant.

Though the Blue Gum, *E. globulus*, is nothing like as reliable, it is a fast-growing, expendable plant for sub-tropical planting on the patio or paved terrace, either in a bed or a container, with outstandingly beautiful, glaucous blue leaves and stems. For this purpose I sow the seed singly in peat pots in August under glass, moving the tiny seedlings on into paper pulp pots which can be torn, so that the roots suffer no disturbance when planted out in May. By the following September this eucalypt should, under normal conditions, achieve a height of 4 to 6 feet. The greatest growth is during August and September.

Gum, Blue (See EUCALYPTUS).

Holly (See ILEX).

Ilex (Holly). Holly does well on light, sand-type soils, though it does not relish cold, stiff soils nor endure winter waterlogging. It has been slow to get away on our thin hungry sand, perhaps because it is difficult to move it with a 'ball', though once it has taken root it goes ahead. It should be planted in May and may need some water its first summer, should the soil be poor. As a hedge there are few plants to equal our common holly, *Ilex aquifolium*, even if it must contend with shade and poverty of soil. I know of holly hedges 20 to 30 feet high, dense right to the ground and still fresh and green after many years. The variegated forms are most beautiful, and are some of the brightest gems of the winter garden. Plant both sexes to ensure red or yellow berries. Good kinds are Golden King and Golden Queen, these being outstanding among gold-variegated hollies, while among the silvers is Handsworth New Silver, a compact variety whose dark green leaves are edged with white.

Juniperus. *Juniperus communis* will be found closely associated with birch or rowan on sandy heathland. It likes plenty of sun and is seldom found where other trees cast dense shade. Junipers are among the most suitable conifers for very calcareous soil, though they grow well on acid sand, as in Surrey. They suffer badly from heavy falls of snow, though their powers of recovery are so great that the tips of old branches which have been laid low soon resemble a fresh group of healthy young trees. While the wild juniper is too tall for most gardens, *J. communis hibernica* grows slowly to 12 feet, making a dark green specimen tree.

Laburnum. The Laburnum, closely allied to the broom, enjoys similar conditions of poor, sandy soil where it flowers most profusely and is deservedly one of the most popular of spring-flowering trees. The common laburnum, *Laburnum anagyroides* is better replaced by the very free-flowering hybrid *vossii* or by the Scotch Laburnum, *L. alpinum* which flowers a couple of weeks later. Both grow to a height of about 15 feet.

May (See CRATAEGUS).

Mountain Ash (See SORBUS).

Myrtle (See MYRTUS).

Myrtus (Myrtle). *Myrtus luma*, the Chilean Tree Myrtle, 8 to 10 feet tall and only reliable in mild localities, bears creamy-white flowers in late summer, black fruits and conspicuous light brown bark which peels in patches, revealing the beautiful cream inner surface. A small avenue is delightful.

Olearia traversii. A rapid grower for sandy gardens which are free from frost, resembling a Lombardy Poplar in shape but only growing to about 15 feet. It has bright green leaves with silvery undersides which flash with every breeze. In its native New Zealand it is largely used for re-claiming sand-dunes. Plant from pots.

Pinus mugo. Though many of the sand-loving pines are too large for small gardens, *Pinus mugo* is ideal for our purpose, always managing to look healthily evergreen, however poor or sandy the soil may be. It has no main leader but forms a bush up to 12 feet tall, fully clothed to soil level with branches, so that it has as wide a spread as it is high. I admired it in the north of Scotland, growing sturdily in the most unpromising conditions of barren sand. If this should be too big, plant instead *P. mugo* Gnome, which grows slowly to 5 or 6 feet.

Robinia (False Acacia). The so-called False Acacia is not planted as freely as it might be, owing to its rather brittle growth, yet it is one of the most beautiful of small trees, both in flower and foliage, and especially suited to hot, dry soils where its growth is less lush than on heavier land

and its brittle shoots more wiry. The Rose Acacia, *Robinia hispida*, and *R. kelseyi*, are among the loveliest of early summer-flowerers, with pendulous pink flowers in June. Give shelter from cold winds.

Rowan (See SORBUS).

Snowy Mespilus (See AMELANCHIER).

Sorbus aucuparia (Mountain Ash or Rowan). The Mountain Ash is too well known to need description and is often found on sandy heath-lands in company with birch and holly. Far better for the modern garden is the elegant *Sorbus hupehensis* from China, with faintly pink berries turning pearly white. Birds do not devour its berries with the same rapaciousness as they do the red fruits of the wild rowan. If you must have red berries plant the smaller *S. scopulina*, which grows slowly to an ultimate height of 12 feet with a spread of 8 feet. Its large handsome leaves are glaucous, and the pillar-box-red berries are carried in heavy clusters. The rowans do best in districts with heavy rainfall.

Weeping Birch (See BETULA).

DWARF AND SLOW-GROWING CONIFERS

There is great demand for slow-growing conifers for the small modern garden, dwarf trees which never outgrow their station but provide immediate interest when the garden is new. They are easy on sandy soil, though they should be well watered in, and also watered again during the first few months after transplanting; do not let them dry out at all. Once they have become established they take very kindly to warm soils such as sand. Look out for them at the now-popular garden centres where many may be seen growing in containers ready for taking home to the garden.

There are beautiful evergreens or evergolds, and some with bright glaucous blue foliage, which enrich the winter garden with colour when there is little else of interest.

Thuja occidentalis Rheingold has fine golden foliage which turns to a blaze of bronze; its effect is quite striking above a carpet of the winter-flowering heaths. *Chamaecyparis lawsoniana fletcheri* has a pyramidal shape of blue-grey feathery foliage. I like to give it a neat trim in the spring. It has taken this dwarf conifer six years to reach 4 feet on our sandy soil. *C. pisifera filifera aurea* is low-spreading with drooping branches of golden-yellow. The Irish Juniper, *Juniperus communis hibernica*, grows like a slender pillar to 8 feet but takes years to achieve it, and *J. chinensis pfitzeriana* spreads horizontally to cover the ground with its dark green foliage.

Countless more of these exciting small conifers may be seen in nurseries which specialise in slow-growing dwarf trees, and anyone who is interested should read *Dwarf Conifers* by H. G. Hillier, published jointly by The Alpine Garden Society and The Scottish Rock Garden Club.

CHAPTER TEN

Bulbous Plants

I HAVE interpreted with some licence the term 'Bulbous Plants', including, besides true bulbs, those with corms and tubers. Probably few things give greater pleasure than the planting of bulbs with the eager expectation of their flowers in due season, and those with well-drained, sandy soil are in a position to gratify this by a judicious selection, and maintain a display over most of the garden year. Bulbs are one of the forerunners of spring, and extend the flowering season when herbaceous plants are finished.

One of the advantages for the amateur gardener is the near-certainty of success, since the flowers are mysteriously already in the bulbs we buy. They are finished products, not dependent so much on the soil as on the conditions which formed them the previous season. Naturalised bulbs, however, will not increase their numbers unless the soil is to their liking. Bulbs like a good, well-drained loam, so that excessively sandy soils should be improved beforehand by the addition of decayed vegetable matter or hop manure—never add fresh animal manures, and do not use peat or leafmould except for the shade-lovers. Bulbs will increase in number if soil and conditions are right—start with a few of each kind until it is discovered which grow best. The collection can later be enlarged.

The range of bulbous plants is so extensive that I have confined myself to those which grow well on sandy soil without too much alteration of its character, omitting many common kinds which prefer a heavier, wetter soil than sand. Snowdrops, for instance, will grow anywhere in shady woodland, though the flowers are larger and more prolific on good soil. It is a mistake to suppose that anemones like a thin, sharply-drained soil; the Wood Anemone prefers a leafy loam and the Poppy Anemone grows best where a sandy soil contains lime and has been enriched with some moisture-retaining material; without this type of material stalks will be short and flowers small. Lilies may be grown on sandy soil, they like good drainage, and one recalls how that great gardener, Miss Jekyll, described the planting of the huge *Lilium giganteum* (now *Cardiocrinum giganteum*) on her Surrey sand, but not without bringing in cart-loads of rotting vegetable matter from outside. A thin, hungry sand is not the natural habitat of lilies, even though its character may be altered to accommodate them.

Here, then, are some bulbous plants which are happy on light sandy soil and will do well on it.

Allium. We should not discard the garden relatives of our common onion because we fear the unpleasant smell. It is rarely apparent in garden forms unless the bulbs are handled, though we should regard with grave suspicion the wild garlic, which is threatening to menace the daffodil farms in the Isles of Scilly. The pretty *Allium moly*, 9 inches tall, with bright golden flowers in June, will naturalise on coastal sands. It is best taken up and replanted at intervals. The tall, imposing *A. rosenbachianum*, with large, light purple umbels on 3-foot stems from May to the end of the season, is a good border plant. One of the best-coloured alliums is *A. caeruleum*. The flowers are a beautiful blue, on firm, 2-foot stems. The Perennial Chives, *A. schoenoprasum*, make a good ornamental edging, with round, rosy-violet flower-heads reminding one of a tall thrift, and the Giant Chives, *A. s. sibiricum*, are a good foil for other border plants. They have parma-violet heads and rushy foliage. The spherical heads of the alliums when cut and dried, before they are spoilt by frost, are much in demand for winter flower arrangements.

Amaryllis. The Belladonna Lily, *Amaryllis belladonna*, carries its pink and white, funnel-shaped flowers on naked stalks during early autumn, its strap-like leaves appearing in spring and dying down.

The main essentials to ensure flowers are a light soil with good drainage and a hot, sun-baked situation such as a south wall provides. Plant deep—6 inches is not too much. Our own bulbs, which flower magnificently, are quite 8 inches in the ground. Water copiously in spring when the leaves appear, and afterwards leave them alone for the sun to do its work. Transplant, where necessary, in June, though the Belladonna Lily is better left undisturbed. A winter mulch should be applied in frosty districts.

Belladonna Lily (See AMARYLLIS).

Bluebell (See ENDYMION).

Bluebell, Spanish (See ENDYMION).

Chionodoxa (Glory of the Snow). The gentian-blue *Chionodoxa sardensis* and the blue and white *C. luciliae*, are delightful in March, and there is a pink form, *C. l.* Pink Giant, and a white, *C. l. alba*. Plant the small bulbs in autumn and do not disturb until they become so crowded they cease to flower well.

Chives (See ALLIUM).

Colchicum. Often mistakenly called the Autumn Crocus, the colchicum has large, crocus-like flowers in September and October, and the corm is a huge tuberous affair, quite unlike the small, neat crocus corm. When planting colchiums, care should be taken in positioning them, for the coarse green leaves which appear in spring are ungainly in the wrong place and must be allowed to die down naturally, or the flowers will be sadly diminished. They may be planted in short grass where the nakedness of the flowers on bare stems is not noticeable, or they look well at the foot of shrubs. They like a sandy soil and increase to make large colonies. The true 'Naked Ladies' is found wild in parts of eastern England and is not to be compared with such kinds as *Colchium speciosum*, or its white form *C. s. album*. *C. s. bornmuelleri* is one of the earliest and the best, and the huge variety Water Lily is outstanding. Colchicums are not cheap, but each enormous bulb yields many flowers even in its first season, and increases freely. Plant in August and you will not have long to wait for the large chalice flowers. Most are some shade of pale lilac.

Crocosmia masonorum. When this South African plant first made its appearance in this country, it was looked upon as doubtfully hardy, but recent severe winters when it came through unscathed in many districts have proved it hardier than was suspected. The intense orange-flame flowers on branching stems, rather like a gargantuan montbretia, brighten the August garden and are also good as cut flowers. One or two are good, though a group is quite outstandingly beautiful. Spread a covering of leaves or bracken in autumn over the crowns of the plants as a protection against frost in a frosty locality.

Crocus. The crocus ranks high among dwarf corms for well-drained, sandy soils in a sunny situation. In grass, at the edge of shrub borders, or in pockets in the rock garden they are most charming. Winter- and early spring-flowering species and their varieties are smaller and more delicate than the later-flowering large Dutch crocuses. Lovely varieties of *Crocus chrysanthus*, many with enchanting bird names, are among the first to flower. Best known of the earlies is perhaps *C. tomasinianus*, known to many as 'Tommy' and its deeper purple form, Whitewell Purple. They naturalise from self-sown seed and offsets, so that they are often regarded on some soils as a weed. Other spring-flowerers are the rich lilac *imperati*, the blue *sieberi* and the golden *susianus*. *C.*

speciosus quickly follows *kotschyanus* (*zonatus*) in the autumn garden, spreading rapidly to make magnificent carpets of colour. Plant spring-flowering crocus in autumn, and autumn-flowering ones in August. The greatest menace is mice, which eat the corms, but a moth ball or two, put in with each group, acts as a deterrent. The large-flowered Dutch hybrids in purple, striped or white may be planted in August to extend the spring display. Crocus need shallow planting and each corm will produce a succession of flowers, opening wide in the sun.

Cuban Lily (See SCILLA).

Daffodil (See NARCISSUS).

Endymion. Our common bluebell, *Endymion non-scriptus*, (*Scilla non-scripta*), is one of the glories of our English oakwoods in spring, but it likes the shade of trees and hates to be dug up and taken home to sunny gardens. Try, instead, some of the good garden forms. *E. hispanicus* (*Scilla hispanica*) is altogether sturdier and a good cut flower, with large bells all round the spike instead of down one side, deep blue or pink or white. It is reliable anywhere, needing no care or attention and will increase rapidly at the edge of shrubberies and is a useful covering for awkward places.

Feather Hyacinth (See MUSCARI).

Flower of the West Wind (See ZEPHYRANTHES).

Galanthus (Snowdrop). The common snowdrop, *Galanthus nivalis*, sheets our woods with white if growing on leafy soil in semi-shade, but on hotter, drier soils such kinds as the large-flowered, grey-leaved *G. elwesii*, the Christmas-flowering *G. byzantinus*, and *G. nivalis cilicicus*, prove more successful. All snowdrops like lime and chalk. They should be transplanted while in full growth as soon as the flowers have faded, as if they are left out of the soil for long they lose vitality.

Galtonia candicans (Cape Hyacinth). Summer-flowering bulbs are often overlooked, though there is no excuse for omitting the strikingly handsome *Galtonia candicans*, with stout 3-foot stems hung with drooping white bells in August. It is cheap to buy and easy to grow. Plant in March in groups, among hardy plants or among shrubs, and only disturb when overcrowding becomes obvious by loss of flower. On extremely sandy soil they repay liberal applications of water during their growing season.

Gladiolus. The most rewarding place for the large-flowered gladioli is in the vegetable garden where the ground can be heavily manured in autumn in readiness for planting from March to May. With copious watering during their growing season they do very well indeed on sandy soils. Do not neglect the small-flowered kinds, many with delicate ruffled petals. On well-drained soil, where the climate is mild, gladioli may be left in the ground to overwinter. The early-flowering *G. byzantinus*, bright magenta, is almost a weed on light sandy soils.

Glory of the Snow (See CHIONODOXA).

Grape Hyacinth (See MUSCARI).

Ipheion uniflorum. This easy and pretty plant should be in every garden. Its white, blue-flushed flowers on 9-inch stalks are one of the harbingers of spring. A shady north-facing aspect suits it best. It has gone under the name *Triteleia uniflora*.

Iris. Whatever the weather, the lovely *Iris histrioides major*, so astonishingly resistant to snow and ice, will flower in January. The stout little fellow does not grow higher than 9 inches, though its flowers are large and a beautiful blue with gold markings. On very sandy soil it is apt to break up into small breeder bulbs and may need lifting every few years. The sweetly scented *I. reticulata*, dark violet, with an orange flash, prefers a limy sand to an acid one, and each year our own small patch has increased in size to give a brilliant display in the February sunshine. Its light blue variety Cantab is more money, but worth it for the contrast it makes with those of deep violet. J. S. Dijt is red-purple.

Though the English irises prefer a moister, heavier soil, the Dutch and Spanish are very good indeed. The Dutch irises flowering in early June are closely followed by the Spanish a couple of weeks later. Bulbs are so cheap it is worth trying out new kinds each year. Plant in an open position in sun.

Ixia. Ixias are not hardy everywhere and in all but the warmest localities should be replanted annually. They are ideal for a poor starved soil in a warm corner. The brilliant flowers are red, flame or yellow. Flower-arrangers should not allow the early-flowering *Ixia viridiflora* to escape their net. Its sea-green flowers, with dark blotches at the throats, are invaluable for early summer arrangements.

Montbretia. The common orange montbretia is only fit for the wild garden where it can spread to form a carpet of pale green leaves with few flowers unless the corms are lifted and separated regularly. Beautiful hybrid forms have larger flowers on long stems and come through normal winters on well-drained, sandy soil. In cold districts they should have the protection of some covering against frost. The delicate *Montbretia rosea*, more correctly known as *Tritonia rosea*, though not so well known in gardens, appears hardier and overwinters well. It likes a sunny situation in the open, where its rather fragile-looking pink flowers on slender, wiry stems can be admired.

Muscari (Grape Hyacinth). *Muscari botryoides*, deep blue, multiplies so rapidly on sandy soil I prefer to plant it in grass, where it spreads a carpet of blue, to offset the yellow daffodils. The rather untidy leaves, as they fade, are a nuisance in the border but hardly noticed on a grassy bank. Do not be content with only blue grape hyacinths, plant the white *M. b. album* as well. The variety *M. armeniacum* Cantab, a lighter blue, flowers a couple of weeks later. Plant in July or August. *M. azureum*, only 4 inches high, and correctly called *Hyacinthus azureus*, is neater in all its parts. Try also the small white form. They are tidy growers and are happy in the company of the scillas and the chionodoxas. *M. comosum monstrosum*, the Feather Hyacinth, always excites interest. It is a dense, loose tuft of soft bluish-violet in May. Catalogues often have it listed as *M. plumosum*.

Narcissus. Though we have a certain success with the daffodils on our sandy soil, there is no doubt they prefer a heavier medium than sand, and on light soil one is very dependent on sufficient rainfall to allow the leaves to die down gradually.

Nerine. The hardiest of these beautiful South African plants is *Nerine bowdenii*. Like the Belladonna Lily, *Amaryllis belladonna*, it produces most of its foliage in the spring; it likes the same sun-baked situation, but unlike the Belladonna, the vivid pink flowers of *N. bowdenii* have a delicacy not to be found in the earlier-flowering bulb. The petals are narrower and more curled; indeed they look so exotic one can scarcely credit they flower in October in the open garden. If you can bear to pick them, they have a long life as cut flowers, no less than two to three weeks. The nerine increases steadily and flowers unfailingly when planted just beneath the surface of the soil; deep planting often accounts for its failure to flower well. It is hardy in so many widely differing climatic conditions that we need have no real fears for its safety, except on cold, waterlogged ground. It will flower exuberantly on pure gravel. If you succeed with *N. bowdenii*, try Fenwick's Variety, with longer stems, flowers of brighter rose-pink, and more of these blooms to each umbel. Bulbs are not cheap to buy but it is worth investing in one or two. Plant in spring.

Scilla. The bright blue *S. sibirica atrocoerulea*, also known as Spring Beauty, is a small gem, a robust variety which is a great improvement on the species. Plant the bulbs in September 3 inches deep and 2 to 3 inches apart. The handsome *S. peruviana* produces rosettes of green leaves from which rise in May large umbels of deep blue flowers, lengthening as they age. It is often called the Cuban Lily though it really comes from Mediterranean regions. It increases steadily on sandy soil, and new plantings can be made by lifting every few years and replanting at once.

Snowdrop (See GALANTHUS).

Tigridia pavonia. The brilliant tigridias flower best in sandy soil in a sunny situation. The flowers, borne on 18-inch stems, are mostly scarlet with yellow mottling, and though they live only for a day, give a dazzling display in August and September. Lift after they have flowered, dry off and replant in April.

Zephyranthes candida. The Flower of the West Wind is a gem for the foot of a sunny wall on light sandy soil where drainage is good. It gets sufficient food to sustain it even in a gravel path, where it will carry the glistening white flowers, rather like white crocuses, from August to October in a good year. When happy it increases freely. Plant in July.

Sand-loving Annuals

THE exceptionally severe winter of 1962-63 did much to encourage an even greater use of annuals then ever before, since they were grown to fill the gaps left by dead shrubs and herbaceous perennial plants, and their popularity has not waned, as we discovered new kinds and rediscovered old friends.

Annuals are notorious sun-lovers and no soil is more suited to their needs than a warm, sandy one. Even the half-hardy zinnias, asters and stocks may safely be sown *in situ* in early May.

In gardens on sand the July and August period is one of the most difficult in the whole gardening year, and if the summer should be a hot one with little rain, there will be a lack of colour unless annuals are planted. Drought is a great shortener of the flowering season of perennial plants but if the ground has been well prepared, many annuals will continue on until the frosts.

Autumn sowing. One of the chief recommendations of a sandy soil is that a large number of annuals may, with advantage, be sown in the open ground in autumn. Anyone who has not had previous experience of autumn-sown annuals will be astonished at their vigour—a single plant can cover more ground than four times as many plants which are sown in spring. They come into bloom earlier, sometimes at the end of May instead of mid-July, and not only are the plants more robust and the individual flowers finer, but the plants have a longer season of flower. The stock-flowered larkspur is very successful, and others are eschscholzia, nemophila, nigella, poppy, cornflower, godetia and clarkia.

The date of sowing should be chosen so that there is no chance of plants flowering before the winter, and yet seedlings must have time to establish themselves before the cold weather. This is a matter of experience in different districts.

On warm, quick-draining soils it is possible to get two distinct seasons of flower on the same ground, by autumn-sowing, and by re-sowing when the autumn-sown annuals have gone out of flower.

Preparation of the ground. Even annuals, which have a short span of life and take little out of the soil, do better where the ground has been properly prepared. If the soil is nothing but pure sand and there is nothing to add to it in the way of humus, stir up the sand and let the air get into it. Annuals which are sown on a heap of sand will be covered with flower and foliage—more flower than foliage since sand is productive of blossom at the expense of leaves—but these same annuals, sown at precisely the same time, on a bed of sand a few inches above a rocky subsoil or over an unbroken hard pan which has not been disturbed, will be spindly and not worth the sowing. Never think because it is pure sand that it does not need deep cultivation at the start. A gardener's battle is half won if begun correctly, and once sandy soil has been deeply trenched it seldom requires trenching again, though nothing in later years will compensate for faulty initial work, as I have found to my cost in our own sandy garden.

Many a newcomer to a sandy plot has been deceived by a thin skin of dark peaty top soil, lying only a few inches above wretched yellow sand. Keep this valuable peaty humus and lay it aside till digging is complete, and replace it in the top surface soil. When the ground has had time to settle, tread it firmly to ensure that no air-pockets prevent the roots of the small seedlings from making contact with the soil.

An early start. In general, the earlier seed is sown the earlier it will germinate and the plants flower. Seed sown in September will probably flower in late May or June; seed sown in March and April will flower from July to September; and seed sown in May will flower from August to October. If you have a greenhouse you can use it to get an early start with the half-hardy annuals; even an airing cupboard may be pressed into use. Antirrhinums do very well indeed when sown here in February or March, and since it is not always appreciated that these will make a fine display on extremely poor, sandy soil that lies so high that most plants perish from want of moisture during their growing season, it would be a mistake not to include them. In such stony, parched situations which tax the ingenuity of the gardener the antirrhinums may be planted with every chance of success; they revel in a place which is fatal to almost everything else. Plant out when still quite small.

Large borders of hardy and half-hardy annuals at the Royal Horticultural Society's Wisley Garden provide excellent examples of what may be achieved on light hungry sand.

My special recommendations would be:

HARDY ANNUALS FOR VERY SANDY SOILS

Alyssum Rosie O'Day
Alyssum Royal Carpet
Calendula
Carnation (Chabaud, Enfant de Nice, camellia-flowered)
Chrysanthemum segetum
Dianthus; Bravo is a good variety
Lavatera Loveliness
Limnanthes douglasii
Linum grandiflorum Venice Red
Phacelia campanularia
Silene pendula
Tropaeolum (Nasturtium)

HALF-HARDY ANNUALS FOR VERY SANDY SOIL

Anagallis linifolia phillipsii
Antirrhinum
Arctotis hybrida
Cleome spinosa
Dimorphotheca aurantiaca Salmon Beauty and Goliath
Echium Blue Bedder
Mesembryanthemum criniflorum
Petunia
Portulaca
Tagetes
Ursinia anethoides
Venidium

37 The cones of *Cedrus atlantica glauca*, the Blue Cedar, contribute considerably to its attractiveness

38 A pink-berried Mountain Ash from China, *Sorbus hupehensis* is a graceful small tree

39 *Crocosmia masonorum* comes from South Africa but is quite hardy provided the crowns are given protection in winter in frosty places

40   *Above:* The Grape Hyacinths multiply very fast on sandy soil, and *Muscari botryoides album*, a white variety, is no exception

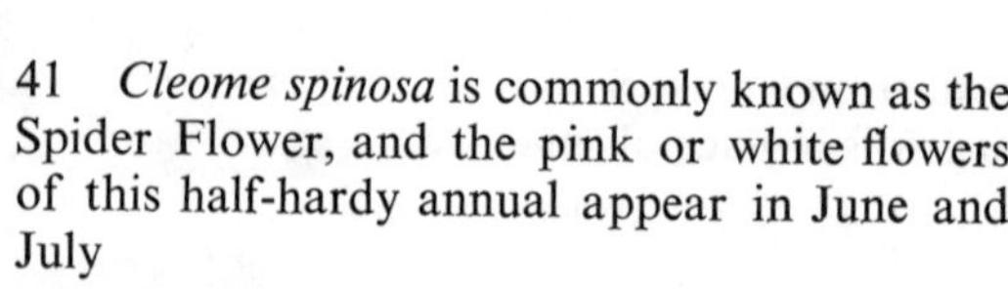

41   *Cleome spinosa* is commonly known as the Spider Flower, and the pink or white flowers of this half-hardy annual appear in June and July

42   *Dianthus* Bravo, a recently introduced annual with brilliant red flowers, is a very good plant for the front of the border

CHAPTER TWELVE

Vegetables

THOUGH it is arguable whether home-grown vegetables are cheaper in the long run than those bought in the shops, it is not disputed that vegetables taken direct from the garden to the pot are infinitely preferable, full of vitamins and flavour.

All soils are improved by the addition of organic material and no soil benefits more than a sandy one. First-class crops will not be produced without it—the vegetable garden should have first priority.

Preparation of soil. At the outset incorporate as much dung, leafmould and compost as you can lay your hands on, and if supplies are limited it is best to concentrate it on a small area, rather than spread it thinly over a larger one, growing the grossest feeders on the part treated, and varying the cropping plan each year, so that in time the whole garden gets its share. Give a good digging at the start, removing all perennial weeds such as bindweed, dock and thistle. Any bulky material laid at the bottom of the trench when the soil is dug will help to bind it together.

Once the initial digging has been completed, give topdressings only —light sand soils need stabilising, not disturbing. With this type of soil anything that is put in moves quickly downwards and if it is put too deep the roots cannot get to it. Lay annual mulches of manure, leafmould or garden compost along the rows of vegetables, for these surface-dressings, laid after rain, are the keynote to success for the vegetable garden on sand, building up a fertile topsoil where roots readily make use of it. In the vegetable garden, where appearance is less important, black polythene is widely used, preventing too rapid evaporation and conserving moisture. Lay it when the ground is moist after a heavy downpour.

Watering. Water often, remembering that during almost the entire period in which vegetables grow best, rainfall is normally at its lowest, using, where possible, modern methods of irrigation to keep the crops on the move. Never wait for rain to do the work for you, for such crops as spinach, radish and lettuce are tough and bitter if their growth is checked by drought. Never listen to those who will tell you that artificial rain will scorch your plants on a bright sunny day. This is not so, though large drops from a watering-can will do so, but if the drops of water are thrown up well into the air, they disintegrate into tiny droplets which wet the surface of the leaves evenly and do not scorch them.

The water required depends largely on the season, the climate and the soil, a sandy one needing twice as much as heavy clay. As soon as one crop is cleared, water the ground before it is cultivated again, and re-sow or plant before the soil has had time to dry out completely.

Planting. On sandy soil, sowing and planting may begin earlier in the year than on heavier land, and plant growth will continue later in the season. This is due to the fact that there is no great volume of water content to be heated, and that these early, warm soils possess such perfect aeration and ventilation, due to their large pores, that there is a constant supply of oxygen to stimulate the living organisms in the soil, at the same time facilitating the dispersal of carbon dioxide which inhibits plant growth. With good management two or even three crops may be grown on the same piece of ground.

The list of vegetables most suited by light soils is extensive but does not include all we may wish to grow, so that I am reminded of a man who drew up a long list of all the things he could not grow, when a friend said to him: 'What you should do is to draw up a short list of the the few things you *can* grow, and grow them really well.'

Artichoke, Jerusalem. This plant prefers a dry, well-drained soil. Plant in March and obtain smooth tubers instead of knobbly ones, which are easier to deal with in the kitchen.

Asparagus. This delicious vegetable likes a sandy, well-manured soil which warms up quickly in the spring. The flavour of asparagus grown on the salty sand-dunes in Holland is said to be particularly delicious. Care should be taken in preparing the bed, all perennial weeds being removed. An asparagus bed may remain for many years if rows are given a topdressing of compost each autumn. Plant one- or two-year-old crowns in late March and do not cut until the second or even the third year. Spray the bed with paraquat weedkiller before the first shoots appear through the soil to kill germinating weed seeds.

Broad Beans. These will grow in almost any soil. Plant longpod types in November, and the Windsor varieties in the spring. Blackfly is the worst pest to be feared, but this can be discouraged by nipping out the tops of the plants when the first flowers appear, and by spraying with liquid derris.

Dwarf French Beans. Few crops are easier on a dry sand than this. Sow when the soil is warm in late April in the south, and in May in the north; and for a succession continue into July, sowing in short rows, for kidney beans should be picked when young.

Runner Beans. These are deep-rooting and need generous applications of dung, decayed vegetable matter, etc. Sow when the soil has warmed up in May and make a second sowing in June. The usual way of growing them is on wire-netting or up bean sticks lashed together at the top with another bean pole. We drive in stout poles, 4 feet apart, and round each pole we sow four to six beans. This method is much practised on the Continent and has the advantage, on shifting, sandy soil, of being more stable in the ground and presenting less front to the wind, which often blows down a row of beans, but is able to blow through poles set apart in this manner.

Beetroot. Though beetroot will grow perfectly well on heavier ground, it is happiest on a light sandy soil which has been previously fed with humus. Sow a couple of seeds every 4 inches, instead of in a continuous row, in mid-April, and again in mid-May.

Beet Seakale (Swiss Chard). This dual-purpose vegetable, of thick, white mid-ribs which can be served as seakale, and dark green leaves as spinach, does well on any soil, but excessively sandy soil should be enriched with humus to retain moisture through the summer. Sow in early May and again at the end of June.

Brussels Sprouts. Brussels sprouts can only be grown well on sandy soil when it has been well fed with old dung or good compost in the autumn. To grow good buttons, the soil must be firm—always difficult on sand. Stake each plant.

Cabbage. Brassicas are difficult to grow on acid sand, suffering badly from club root, which is largely due to acidity. A plentiful use of lime is necessary to overcome it. On our alkaline sand there is no such difficulty, so long as the fertility of the soil is maintained. Spring cabbage is excellent, since it matures when the ground has not completely dried out after the winter rains. Savoys are not so good, growing when the rainfall is normally at its lowest.

Carrot. The best carrots are grown on deeply dug sandy soils. No manure should be added for carrots, though where the soil is in poor condition apply a general fertiliser a few days before sowing. Early carrots require a soil that warms up quickly, and may be sown in sheltered spots in March as soon as the tilth is right. Continue sowing until mid-May and sow an early carrot again in mid-July, for pulling in the autumn. The pelted carrot seed is excellent; no thinning is required, as each coated seed may be sown separately and there is less danger of carrot fly.

Kale. This is a useful crop that survives the severest winters and produces plenty of greens during January and February. It is scarcely ever attacked by club root and will grow on soils too poor for others of the brassica family. Plant on ground that has borne early potatoes. Try the asparagus kale.

Leeks. Leeks grow on almost any soil but excessively poor sandy soil will not produce the best crops unless it is fed. Leeks need moisture. Plant each small leek in holes 6 inches apart; water, and leave the holes open, for the watering will wash sufficient soil down to cover the roots.

Lettuce. Sandy soils should be enriched with rotted manure or good compost to grow crisp, quick-maturing lettuces. They must not suffer any check and frequent watering is essential during drought. On light soils, if the right varieties are grown, it is possible to have lettuce from early May until late autumn and the season may be prolonged by early and late sowings under cloches. For general purpose I have not found a better lettuce for very sandy soil than Baker's All Heart, and an American variety, 'Cornel 456, is particularly good on sand.

Onion. Excellent onions may be grown on light sandy soils which have been deeply dug and manured the previous autumn. They like a firm bed in an open site and very sandy soils may be rolled or beaten down with a spade after sowing. Onion sets are easier and less subject to onion fly. Plant the bulblets at the end of March, 6 inches apart, so that only the necks are showing, and should they come up out of the soil push them firmly back.

Parsnips. Parsnips grow well on light soils or heavy clay, the great disadvantage being that they occupy ground for almost the whole year. Sow in February on a piece of land which has been previously manured; never manure specially for parsnips, or, like carrots, they will fork.

Peas. A few weeks before sowing, dig compost or well-rotted manure into the proposed drills, and for the first sowing in February or early March choose a sunny site on well-drained soil. Leave the drill about 3 inches below the soil level, for this will help to trap the moisture, without which good peas cannot be grown. Choose early varieties for the first sowing, main-crop for April and May, late varieties for June, and round-seeded for autumn sowing.

Potatoes. Sandy soils should be heavily manured in autumn and left rough till late March or early April, when the tubers are planted $1\frac{1}{2}$ feet apart in rows 2 feet apart. Potatoes like manure but are averse to lime.

Earth up the plants when 8 inches high, leaving the top 6 inches uncovered and earth again a month later and once more after that.

Purple-sprouting Broccoli. Often called the 'poor man's asparagus', this is an excellent vegetable for sandy soils for the pre-Christmas period or early spring when vegetables are scarce. Sow in April or early May and plant out on firm ground 2 feet apart.

Radish. This is a crop which must be grown without check. The plants need a little fine organic matter in the top few inches of a sandy soil, to prevent loss of moisture which will render them tough and bitter. Moistened sedge peat, spread along the drills before sowing is excellent.

Seakale. The vegetable we cultivate in gardens was derived from the wild *Crambe maritima,* which occurs in the British Isles on sands near the sea from Fife southwards. The vegetable we use is cultivated for its leaf stalks, usually grown in darkness to force them into growth during the winter months. The blanched leaf stalks should be cut when young and tender. Seakale grows best on light sandy soil with a fairly high lime content which has been enriched with rotted organic material. Buy prepared root cuttings and plant in spring. They can later be covered with seakale pots or boxes, as for forced rhubarb, and surrounded with fresh manure or straw.

Spinach Beet. This is a good substitute for the annual spinach, which needs a moister soil than sand provides. To have a supply of perpetual spinach all the year, sow in April and again in August. Keep watered if the weather is dry.

Index

Abbreviation: p = illustration facing page given

ALPHABETS AND ORNAMENTS, Ernst Lehner. Well-known pictorial source for decorative alphabets, script examples, cartouches, frames, decorative title pages, calligraphic initials, borders, similar material. 14th to 19th century, mostly European. Useful in almost any graphic arts designing, varied styles. 750 illustrations. 256pp. 7 x 10.   21905-4 Paperbound $4.00

PAINTING: A CREATIVE APPROACH, Norman Colquhoun. For the beginner simple guide provides an instructive approach to painting: major stumbling blocks for beginner; overcoming them, technical points; paints and pigments; oil painting; watercolor and other media and color. New section on "plastic" paints. Glossary. Formerly *Paint Your Own Pictures*. 221pp.   22000-1 Paperbound $1.75

THE ENJOYMENT AND USE OF COLOR, Walter Sargent. Explanation of the relations between colors themselves and between colors in nature and art, including hundreds of little-known facts about color values, intensities, effects of high and low illumination, complementary colors. Many practical hints for painters, references to great masters. 7 color plates, 29 illustrations. x + 274pp.
20944-X Paperbound $3.00

THE NOTEBOOKS OF LEONARDO DA VINCI, compiled and edited by Jean Paul Richter. 1566 extracts from original manuscripts reveal the full range of Leonardo's versatile genius: all his writings on painting, sculpture, architecture, anatomy, astronomy, geography, topography, physiology, mining, music, etc., in both Italian and English, with 186 plates of manuscript pages and more than 500 additional drawings. Includes studies for the Last Supper, the lost Sforza monument, and other works. Total of xlvii + 866pp. 7⅞ x 10¾.
22572-0, 22573-9 Two volumes, Paperbound $12.00

MONTGOMERY WARD CATALOGUE OF 1895. Tea gowns, yards of flannel and pillow-case lace, stereoscopes, books of gospel hymns, the New Improved Singer Sewing Machine, side saddles, milk skimmers, straight-edged razors, high-button shoes, spittoons, and on and on . . . listing some 25,000 items, practically all illustrated. Essential to the shoppers of the 1890's, it is our truest record of the spirit of the period. Unaltered reprint of Issue No. 57, Spring and Summer 1895. Introduction by Boris Emmet. Innumerable illustrations. xiii + 624pp. 8½ x 11⅝.
22377-9 Paperbound $8.50

THE CRYSTAL PALACE EXHIBITION ILLUSTRATED CATALOGUE (LONDON, 1851). One of the wonders of the modern world—the Crystal Palace Exhibition in which all the nations of the civilized world exhibited their achievements in the arts and sciences—presented in an equally important illustrated catalogue. More than 1700 items pictured with accompanying text—ceramics, textiles, cast-iron work, carpets, pianos, sleds, razors, wall-papers, billiard tables, beehives, silverware and hundreds of other artifacts—represent the focal point of Victorian culture in the Western World. Probably the largest collection of Victorian decorative art ever assembled— indispensable for antiquarians and designers. Unabridged republication of the Art-Journal Catalogue of the Great Exhibition of 1851, with all terminal essays. New introduction by John Gloag, F.S.A. xxxiv + 426pp. 9 x 12.
22503-8 Paperbound $5.00

A HISTORY OF COSTUME, Carl Köhler. Definitive history, based on surviving pieces of clothing primarily, and paintings, statues, etc. secondarily. Highly readable text, supplemented by 594 illustrations of costumes of the ancient Mediterranean peoples, Greece and Rome, the Teutonic prehistoric period; costumes of the Middle Ages, Renaissance, Baroque, 18th and 19th centuries. Clear, measured patterns are provided for many clothing articles. Approach is practical throughout. Enlarged by Emma von Sichart. 464pp.   21030-8 Paperbound $3.50

ORIENTAL RUGS, ANTIQUE AND MODERN, Walter A. Hawley. A complete and authoritative treatise on the Oriental rug—where they are made, by whom and how, designs and symbols, characteristics in detail of the six major groups, how to distinguish them and how to buy them. Detailed technical data is provided on periods, weaves, warps, wefts, textures, sides, ends and knots, although no technical background is required for an understanding. 11 color plates, 80 halftones, 4 maps. vi + 320pp. 6⅛ x 9⅛.   22366-3 Paperbound $5.00

TEN BOOKS ON ARCHITECTURE, Vitruvius. By any standards the most important book on architecture ever written. Early Roman discussion of aesthetics of building, construction methods, orders, sites, and every other aspect of architecture has inspired, instructed architecture for about 2,000 years. Stands behind Palladio, Michelangelo, Bramante, Wren, countless others. Definitive Morris H. Morgan translation. 68 illustrations. xii + 331pp.   20645-9 Paperbound .$3.00

THE FOUR BOOKS OF ARCHITECTURE, Andrea Palladio. Translated into every major Western European language in the two centuries following its publication in 1570, this has been one of the most influential books in the history of architecture. Complete reprint of the 1738 Isaac Ware edition. New introduction by Adolf Placzek, Columbia Univ. 216 plates. xxii + 110pp. of text. 9½ x 12¾.
21308-0 Clothbound $12.50

STICKS AND STONES: A STUDY OF AMERICAN ARCHITECTURE AND CIVILIZATION, Lewis Mumford. One of the great classics of American cultural history. American architecture from the medieval-inspired earliest forms to the early 20th century; evolution of structure and style, and reciprocal influences on environment. 21 photographic illustrations. 238pp.   20202-X Paperbound $2.00

THE AMERICAN BUILDER'S COMPANION, Asher Benjamin. The most widely used early 19th century architectural style and source book, for colonial up into Greek Revival periods. Extensive development of geometry of carpentering, construction of sashes, frames, doors, stairs; plans and elevations of domestic and other buildings. Hundreds of thousands of houses were built according to this book, now invaluable to historians, architects, restorers, etc. 1827 edition. 59 plates. 114pp. 7⅞ x 10¾.
22236-5 Paperbound $4.00

DUTCH HOUSES IN THE HUDSON VALLEY BEFORE 1776, Helen Wilkinson Reynolds. The standard survey of the Dutch colonial house and outbuildings, with constructional features, decoration, and local history associated with individual homesteads. Introduction by Franklin D. Roosevelt. Map. 150 illustrations. 469pp. 6⅝ x 9¼.   21469-9 Paperbound $5.00

THE ARCHITECTURE OF COUNTRY HOUSES, Andrew J. Downing. Together with Vaux's *Villas and Cottages* this is the basic book for Hudson River Gothic architecture of the middle Victorian period. Full, sound discussions of general aspects of housing, architecture, style, decoration, furnishing, together with scores of detailed house plans, illustrations of specific buildings, accompanied by full text. Perhaps the most influential single American architectural book. 1850 edition. Introduction by J. Stewart Johnson. 321 figures, 34 architectural designs. xvi + 560pp.

22003-6 Paperbound $5.00

LOST EXAMPLES OF COLONIAL ARCHITECTURE, John Mead Howells. Full-page photographs of buildings that have disappeared or been so altered as to be denatured, including many designed by major early American architects. 245 plates. xvii + 248pp. $7\frac{7}{8}$ x $10\frac{3}{4}$.

21143-6 Paperbound $3.50

DOMESTIC ARCHITECTURE OF THE AMERICAN COLONIES AND OF THE EARLY REPUBLIC, Fiske Kimball. Foremost architect and restorer of Williamsburg and Monticello covers nearly 200 homes between 1620-1825. Architectural details, construction, style features, special fixtures, floor plans, etc. Generally considered finest work in its area. 219 illustrations of houses, doorways, windows, capital mantels. xx + 314pp. $7\frac{7}{8}$ x $10\frac{3}{4}$.

21743-4 Paperbound $4.00

EARLY AMERICAN ROOMS: 1650-1858, edited by Russell Hawes Kettell. Tour of 12 rooms, each representative of a different era in American history and each furnished, decorated, designed and occupied in the style of the era. 72 plans and elevations, 8-page color section, etc., show fabrics, wall papers, arrangements, etc. Full descriptive text. xvii + 200pp. of text. $8\frac{3}{8}$ x $11\frac{1}{4}$.

21633-0 Paperbound $5.00

THE FITZWILLIAM VIRGINAL BOOK, edited by J. Fuller Maitland and W. B. Squire. Full modern printing of famous early 17th-century ms. volume of 300 works by Morley, Byrd, Bull, Gibbons, etc. For piano or other modern keyboard instrument; easy to read format. xxxvi + 938pp. $8\frac{3}{8}$ x 11.

21068-5, 21069-3 Two volumes, Paperbound $12.00

KEYBOARD MUSIC, Johann Sebastian Bach. Bach Gesellschaft edition. A rich selection of Bach's masterpieces for the harpsichord: the six English Suites, six French Suites, the six Partitas (Clavierübung part I), the Goldberg Variations (Clavierübung part IV), the fifteen Two-Part Inventions and the fifteen Three-Part Sinfonias. Clearly reproduced on large sheets with ample margins; eminently playable. vi + 312pp. $8\frac{1}{8}$ x 11.

22360-4 Paperbound $5.00

THE MUSIC OF BACH: AN INTRODUCTION, Charles Sanford Terry. A fine, nontechnical introduction to Bach's music, both instrumental and vocal. Covers organ music, chamber music, passion music, other types. Analyzes themes, developments, innovations. x + 114pp.

21075-8 Paperbound $1.95

BEETHOVEN AND HIS NINE SYMPHONIES, Sir George Grove. Noted British musicologist provides best history, analysis, commentary on symphonies. Very thorough, rigorously accurate; necessary to both advanced student and amateur music lover. 436 musical passages. vii + 407 pp.

20334-4 Paperbound $4.00

JOHANN SEBASTIAN BACH, Philipp Spitta. One of the great classics of musicology, this definitive analysis of Bach's music (and life) has never been surpassed. Lucid, nontechnical analyses of hundreds of pieces (30 pages devoted to St. Matthew Passion, 26 to B Minor Mass). Also includes major analysis of 18th-century music. 450 musical examples. 40-page musical supplement. Total of xx + 1799pp.
(EUK) 22278-0, 22279-9 Two volumes, Clothbound $25.00

MOZART AND HIS PIANO CONCERTOS, Cuthbert Girdlestone. The only full-length study of an important area of Mozart's creativity. Provides detailed analyses of all 23 concertos, traces inspirational sources. 417 musical examples. Second edition. 509pp.
21271-8 Paperbound $4.50

THE PERFECT WAGNERITE: A COMMENTARY ON THE NIBLUNG'S RING, George Bernard Shaw. Brilliant and still relevant criticism in remarkable essays on Wagner's Ring cycle, Shaw's ideas on political and social ideology behind the plots, role of Leitmotifs, vocal requisites, etc. Prefaces. xxi + 136pp.
(USO) 21707-8 Paperbound $1.75

DON GIOVANNI, W. A. Mozart. Complete libretto, modern English translation; biographies of composer and librettist; accounts of early performances and critical reaction. Lavishly illustrated. All the material you need to understand and appreciate this great work. Dover Opera Guide and Libretto Series; translated and introduced by Ellen Bleiler. 92 illustrations. 209pp.
21134-7 Paperbound $2.00

BASIC ELECTRICITY, U. S. Bureau of Naval Personel. Originally a training course, best non-technical coverage of basic theory of electricity and its applications. Fundamental concepts, batteries, circuits, conductors and wiring techniques, AC and DC, inductance and capacitance, generators, motors, transformers, magnetic amplifiers, synchros, servomechanisms, etc. Also covers blue-prints, electrical diagrams, etc. Many questions, with answers. 349 illustrations. x + 448pp. 6½ x 9¼.
20973-3 Paperbound $3.50

REPRODUCTION OF SOUND, Edgar Villchur. Thorough coverage for laymen of high fidelity systems, reproducing systems in general, needles, amplifiers, preamps, loudspeakers, feedback, explaining physical background. "A rare talent for making technicalities vividly comprehensible," R. Darrell, *High Fidelity*. 69 figures. iv + 92pp.
21515-6 Paperbound $1.35

HEAR ME TALKIN' TO YA: THE STORY OF JAZZ AS TOLD BY THE MEN WHO MADE IT, Nat Shapiro and Nat Hentoff. Louis Armstrong, Fats Waller, Jo Jones, Clarence Williams, Billy Holiday, Duke Ellington, Jelly Roll Morton and dozens of other jazz greats tell how it was in Chicago's South Side, New Orleans, depression Harlem and the modern West Coast as jazz was born and grew. xvi + 429pp.
21726-4 Paperbound $3.95

FABLES OF AESOP, translated by Sir Roger L'Estrange. A reproduction of the very rare 1931 Paris edition; a selection of the most interesting fables, together with 50 imaginative drawings by Alexander Calder. v + 128pp. 6½x9¼.
21780-9 Paperbound $1.50

AGAINST THE GRAIN (A REBOURS), Joris K. Huysmans. Filled with weird images, evidences of a bizarre imagination, exotic experiments with hallucinatory drugs, rich tastes and smells and the diversions of its sybarite hero Duc Jean des Esseintes, this classic novel pushed 19th-century literary decadence to its limits. Full unabridged edition. Do not confuse this with abridged editions generally sold. Introduction by Havelock Ellis. xlix + 206pp. 22190-3 Paperbound $2.50

VARIORUM SHAKESPEARE: HAMLET. Edited by Horace H. Furness; a landmark of American scholarship. Exhaustive footnotes and appendices treat all doubtful words and phrases, as well as suggested critical emendations throughout the play's history. First volume contains editor's own text, collated with all Quartos and Folios. Second volume contains full first Quarto, translations of Shakespeare's sources (Belleforest, and Saxo Grammaticus), Der Bestrafte Brudermord, and many essays on critical and historical points of interest by major authorities of past and present. Includes details of staging and costuming over the years. By far the best edition available for serious students of Shakespeare. Total of xx + 905pp. 21004-9, 21005-7, 2 volumes, Paperbound $7.00

A LIFE OF WILLIAM SHAKESPEARE, Sir Sidney Lee. This is the standard life of Shakespeare, summarizing everything known about Shakespeare and his plays. Incredibly rich in material, broad in coverage, clear and judicious, it has served thousands as the best introduction to Shakespeare. 1931 edition. 9 plates. xxix + 792pp. 21967-4 Paperbound $4.50

MASTERS OF THE DRAMA, John Gassner. Most comprehensive history of the drama in print, covering every tradition from Greeks to modern Europe and America, including India, Far East, etc. Covers more than 800 dramatists, 2000 plays, with biographical material, plot summaries, theatre history, criticism, etc. "Best of its kind in English," *New Republic.* 77 illustrations. xxii + 890pp. 20100-7 Clothbound $10.00

THE EVOLUTION OF THE ENGLISH LANGUAGE, George McKnight. The growth of English, from the 14th century to the present. Unusual, non-technical account presents basic information in very interesting form: sound shifts, change in grammar and syntax, vocabulary growth, similar topics. Abundantly illustrated with quotations. Formerly *Modern English in the Making.* xii + 590pp. 21932-1 Paperbound $3.50

AN ETYMOLOGICAL DICTIONARY OF MODERN ENGLISH, Ernest Weekley. Fullest, richest work of its sort, by foremost British lexicographer. Detailed word histories, including many colloquial and archaic words; extensive quotations. Do not confuse this with the Concise Etymological Dictionary, which is much abridged. Total of xxvii + 830pp. 6½ x 9¼. 21873-2, 21874-0 Two volumes, Paperbound $7.90

FLATLAND: A ROMANCE OF MANY DIMENSIONS, E. A. Abbott. Classic of science-fiction explores ramifications of life in a two-dimensional world, and what happens when a three-dimensional being intrudes. Amusing reading, but also useful as introduction to thought about hyperspace. Introduction by Banesh Hoffmann. 16 illustrations. xx + 103pp. 20001-9 Paperbound $1.00

POEMS OF ANNE BRADSTREET, edited with an introduction by Robert Hutchinson. A new selection of poems by America's first poet and perhaps the first significant woman poet in the English language. 48 poems display her development in works of considerable variety—love poems, domestic poems, religious meditations, formal elegies, "quaternions," etc. Notes, bibliography. viii + 222pp.

22160-1 Paperbound $2.50

THREE GOTHIC NOVELS: THE CASTLE OF OTRANTO BY HORACE WALPOLE; VATHEK BY WILLIAM BECKFORD; THE VAMPYRE BY JOHN POLIDORI, WITH FRAGMENT OF A NOVEL BY LORD BYRON, edited by E. F. Bleiler. The first Gothic novel, by Walpole; the finest Oriental tale in English, by Beckford; powerful Romantic supernatural story in versions by Polidori and Byron. All extremely important in history of literature; all still exciting, packed with supernatural thrills, ghosts, haunted castles, magic, etc. xl + 291pp.

21232-7 Paperbound $3.00

THE BEST TALES OF HOFFMANN, E. T. A. Hoffmann. 10 of Hoffmann's most important stories, in modern re-editings of standard translations: Nutcracker and the King of Mice, Signor Formica, Automata, The Sandman, Rath Krespel, The Golden Flowerpot, Master Martin the Cooper, The Mines of Falun, The King's Betrothed, A New Year's Eve Adventure. 7 illustrations by Hoffmann. Edited by E. F. Bleiler. xxxix + 419pp. 21793-0 Paperbound $3.00

GHOST AND HORROR STORIES OF AMBROSE BIERCE, Ambrose Bierce. 23 strikingly modern stories of the horrors latent in the human mind: The Eyes of the Panther, The Damned Thing, An Occurrence at Owl Creek Bridge, An Inhabitant of Carcosa, etc., plus the dream-essay, Visions of the Night. Edited by E. F. Bleiler. xxii + 199pp. 20767-6 Paperbound $2.00

BEST GHOST STORIES OF J. S. LeFANU, J. Sheridan LeFanu. Finest stories by Victorian master often considered greatest supernatural writer of all. Carmilla, Green Tea, The Haunted Baronet, The Familiar, and 12 others. Most never before available in the U. S. A. Edited by E. F. Bleiler. 8 illustrations from Victorian publications. xvii + 467pp. 20415-4 Paperbound $3.00

MATHEMATICAL FOUNDATIONS OF INFORMATION THEORY, A. I. Khinchin. Comprehensive introduction to work of Shannon, McMillan, Feinstein and Khinchin, placing these investigations on a rigorous mathematical basis. Covers entropy concept in probability theory, uniqueness theorem, Shannon's inequality, ergodic sources, the E property, martingale concept, noise, Feinstein's fundamental lemma, Shanon's first and second theorems. Translated by R. A. Silverman and M. D. Friedman. iii + 120pp. 60434-9 Paperbound $2.00

SEVEN SCIENCE FICTION NOVELS, H. G. Wells. The standard collection of the great novels. Complete, unabridged. *First Men in the Moon, Island of Dr. Moreau, War of the Worlds, Food of the Gods, Invisible Man, Time Machine, In the Days of the Comet.* Not only science fiction fans, but every educated person owes it to himself to read these novels. 1015pp. (USO) 20264-X Clothbound $6.00

LAST AND FIRST MEN AND STAR MAKER, TWO SCIENCE FICTION NOVELS, Olaf Stapledon. Greatest future histories in science fiction. In the first, human intelligence is the "hero," through strange paths of evolution, interplanetary invasions, incredible technologies, near extinctions and reemergences. Star Maker describes the quest of a band of star rovers for intelligence itself, through time and space: weird inhuman civilizations, crustacean minds, symbiotic worlds, etc. Complete, unabridged. v + 438pp. (USO) 21962-3 Paperbound $3.00

THREE PROPHETIC NOVELS, H. G. WELLS. Stages of a consistently planned future for mankind. *When the Sleeper Wakes,* and *A Story of the Days to Come,* anticipate *Brave New World* and *1984,* in the 21st Century; *The Time Machine,* only complete version in print, shows farther future and the end of mankind. All show Wells's greatest gifts as storyteller and novelist. Edited by E. F. Bleiler. x + 335pp. (USO) 20605-X Paperbound $3.00

THE DEVIL'S DICTIONARY, Ambrose Bierce. America's own Oscar Wilde— Ambrose Bierce—offers his barbed iconoclastic wisdom in over 1,000 definitions hailed by H. L. Mencken as "some of the most gorgeous witticisms in the English language." 145pp. 20487-1 Paperbound $1.50

MAX AND MORITZ, Wilhelm Busch. Great children's classic, father of comic strip, of two bad boys, Max and Moritz. Also Ker and Plunk (Plisch und Plumm), Cat and Mouse, Deceitful Henry, Ice-Peter, The Boy and the Pipe, and five other pieces. Original German, with English translation. Edited by H. Arthur Klein; translations by various hands and H. Arthur Klein. vi + 216pp.
20181-3 Paperbound $2.00

PIGS IS PIGS AND OTHER FAVORITES, Ellis Parker Butler. The title story is one of the best humor short stories, as Mike Flannery obfuscates biology and English. Also included, That Pup of Murchison's, The Great American Pie Company, and Perkins of Portland. 14 illustrations. v + 109pp. 21532-6 Paperbound $1.50

THE PETERKIN PAPERS, Lucretia P. Hale. It takes genius to be as stupidly mad as the Peterkins, as they decide to become wise, celebrate the "Fourth," keep a cow, and otherwise strain the resources of the Lady from Philadelphia. Basic book of American humor. 153 illustrations. 219pp. 20794-3 Paperbound $2.00

PERRAULT'S FAIRY TALES, translated by A. E. Johnson and S. R. Littlewood, with 34 full-page illustrations by Gustave Doré. All the original Perrault stories— Cinderella, Sleeping Beauty, Bluebeard, Little Red Riding Hood, Puss in Boots, Tom Thumb, etc.—with their witty verse morals and the magnificent illustrations of Doré. One of the five or six great books of European fairy tales. viii + 117pp. 8⅛ x 11. 22311-6 Paperbound $2.00

OLD HUNGARIAN FAIRY TALES, Baroness Orczy. Favorites translated and adapted by author of the *Scarlet Pimpernel.* Eight fairy tales include "The Suitors of Princess Fire-Fly," "The Twin Hunchbacks," "Mr. Cuttlefish's Love Story," and "The Enchanted Cat." This little volume of magic and adventure will captivate children as it has for generations. 90 drawings by Montagu Barstow. 96pp.
(USO) 22293-4 Paperbound $1.95

THE RED FAIRY BOOK, Andrew Lang. Lang's color fairy books have long been children's favorites. This volume includes Rapunzel, Jack and the Bean-stalk and 35 other stories, familiar and unfamiliar. 4 plates, 93 illustrations x + 367pp.
21673-X Paperbound $2.50

THE BLUE FAIRY BOOK, Andrew Lang. Lang's tales come from all countries and all times. Here are 37 tales from Grimm, the Arabian Nights, Greek Mythology, and other fascinating sources. 8 plates, 130 illustrations. xi + 390pp.
21437-0 Paperbound $2.75

HOUSEHOLD STORIES BY THE BROTHERS GRIMM. Classic English-language edition of the well-known tales — Rumpelstiltskin, Snow White, Hansel and Gretel, The Twelve Brothers, Faithful John, Rapunzel, Tom Thumb (52 stories in all). Translated into simple, straightforward English by Lucy Crane. Ornamented with headpieces, vignettes, elaborate decorative initials and a dozen full-page illustrations by Walter Crane. x + 269pp.
21080-4 Paperbound **$2.00**

THE MERRY ADVENTURES OF ROBIN HOOD, Howard Pyle. The finest modern versions of the traditional ballads and tales about the great English outlaw. Howard Pyle's complete prose version, with every word, every illustration of the first edition. Do not confuse this facsimile of the original (1883) with modern editions that change text or illustrations. 23 plates plus many page decorations. xxii + 296pp.
22043-5 Paperbound $2.75

THE STORY OF KING ARTHUR AND HIS KNIGHTS, Howard Pyle. The finest children's version of the life of King Arthur; brilliantly retold by Pyle, with 48 of his most imaginative illustrations. xviii + 313pp. 6⅛ x 9¼.
21445-1 Paperbound $2.50

THE WONDERFUL WIZARD OF OZ, L. Frank Baum. America's finest children's book in facsimile of first edition with all Denslow illustrations in full color. The edition a child should have. Introduction by Martin Gardner. 23 color plates, scores of drawings. iv + 267pp.
20691-2 Paperbound $3.50

THE MARVELOUS LAND OF OZ, L. Frank Baum. The second Oz book, every bit as imaginative as the Wizard. The hero is a boy named Tip, but the Scarecrow and the Tin Woodman are back, as is the Oz magic. 16 color plates, 120 drawings by John R. Neill. 287pp.
20692-0 Paperbound $2.50

THE MAGICAL MONARCH OF MO, L. Frank Baum. Remarkable adventures in a land even stranger than Oz. The best of Baum's books not in the Oz series. 15 color plates and dozens of drawings by Frank Verbeck. xviii + 237pp.
21892-9 Paperbound $2.25

THE BAD CHILD'S BOOK OF BEASTS, MORE BEASTS FOR WORSE CHILDREN, A MORAL ALPHABET, Hilaire Belloc. Three complete humor classics in one volume. Be kind to the frog, and do not call him names . . . and 28 other whimsical animals. Familiar favorites and some not so well known. Illustrated by Basil Blackwell. 156pp.
(USO) 20749-8 Paperbound $1.50

EAST O' THE SUN AND WEST O' THE MOON, George W. Dasent. Considered the best of all translations of these Norwegian folk tales, this collection has been enjoyed by generations of children (and folklorists too). Includes True and Untrue, Why the Sea is Salt, East O' the Sun and West O' the Moon, Why the Bear is Stumpy-Tailed, Boots and the Troll, The Cock and the Hen, Rich Peter the Pedlar, and 52 more. The only edition with all 59 tales. 77 illustrations by Erik Werenskiold and Theodor Kittelsen. xv + 418pp. 22521-6 Paperbound $3.50

GOOPS AND HOW TO BE THEM, Gelett Burgess. Classic of tongue-in-cheek humor, masquerading as etiquette book. 87 verses, twice as many cartoons, show mischievous Goops as they demonstrate to children virtues of table manners, neatness, courtesy, etc. Favorite for generations. viii + 88pp. 6½ x 9¼.
22233-0 Paperbound $1.50

ALICE'S ADVENTURES UNDER GROUND, Lewis Carroll. The first version, quite different from the final *Alice in Wonderland,* printed out by Carroll himself with his own illustrations. Complete facsimile of the "million dollar" manuscript Carroll gave to Alice Liddell in 1864. Introduction by Martin Gardner. viii + 96pp. Title and dedication pages in color. 21482-6 Paperbound $1.25

THE BROWNIES, THEIR BOOK, Palmer Cox. Small as mice, cunning as foxes, exuberant and full of mischief, the Brownies go to the zoo, toy shop, seashore, circus, etc., in 24 verse adventures and 266 illustrations. Long a favorite, since their first appearance in St. Nicholas Magazine. xi + 144pp. 6⅝ x 9¼.
21265-3 Paperbound $1.75

SONGS OF CHILDHOOD, Walter De La Mare. Published (under the pseudonym Walter Ramal) when De La Mare was only 29, this charming collection has long been a favorite children's book. A facsimile of the first edition in paper, the 47 poems capture the simplicity of the nursery rhyme and the ballad, including such lyrics as I Met Eve, Tartary, The Silver Penny. vii + 106pp. (USO) 21972-0 Paperbound $1.25

THE COMPLETE NONSENSE OF EDWARD LEAR, Edward Lear. The finest 19th-century humorist-cartoonist in full: all nonsense limericks, zany alphabets, Owl and Pussycat, songs, nonsense botany, and more than 500 illustrations by Lear himself. Edited by Holbrook Jackson. xxix + 287pp. (USO) 20167-8 Paperbound $2.00

BILLY WHISKERS: THE AUTOBIOGRAPHY OF A GOAT, Frances Trego Montgomery. A favorite of children since the early 20th century, here are the escapades of that rambunctious, irresistible and mischievous goat—Billy Whiskers. Much in the spirit of *Peck's Bad Boy,* this is a book that children never tire of reading or hearing. All the original familiar illustrations by W. H. Fry are included: 6 color plates, 18 black and white drawings. 159pp. 22345-0 Paperbound $2.00

MOTHER GOOSE MELODIES. Faithful republication of the fabulously rare Munroe and Francis "copyright 1833" Boston edition—the most important Mother Goose collection, usually referred to as the "original." Familiar rhymes plus many rare ones, with wonderful old woodcut illustrations. Edited by E. F. Bleiler. 128pp. 4½ x 6⅜. 22577-1 Paperbound $1.00

TWO LITTLE SAVAGES; BEING THE ADVENTURES OF TWO BOYS WHO LIVED AS INDIANS AND WHAT THEY LEARNED, Ernest Thompson Seton. Great classic of nature and boyhood provides a vast range of woodlore in most palatable form, a genuinely entertaining story. Two farm boys build a teepee in woods and live in it for a month, working out Indian solutions to living problems, star lore, birds and animals, plants, etc. 293 illustrations. vii + 286pp.

20985-7 Paperbound $2.50

PETER PIPER'S PRACTICAL PRINCIPLES OF PLAIN & PERFECT PRONUNCIATION. Alliterative jingles and tongue-twisters of surprising charm, that made their first appearance in America about 1830. Republished in full with the spirited woodcut illustrations from this earliest American edition. 32pp. $4\frac{1}{2}$ x $6\frac{3}{8}$.

22560-7 Paperbound $1.00

SCIENCE EXPERIMENTS AND AMUSEMENTS FOR CHILDREN, Charles Vivian. 73 easy experiments, requiring only materials found at home or easily available, such as candles, coins, steel wool, etc.; illustrate basic phenomena like vacuum, simple chemical reaction, etc. All safe. Modern, well-planned. Formerly *Science Games for Children*. 102 photos, numerous drawings. 96pp. $6\frac{1}{8}$ x $9\frac{1}{4}$.

21856-2 Paperbound $1.25

AN INTRODUCTION TO CHESS MOVES AND TACTICS SIMPLY EXPLAINED, Leonard Barden. Informal intermediate introduction, quite strong in explaining reasons for moves. Covers basic material, tactics, important openings, traps, positional play in middle game, end game. Attempts to isolate patterns and recurrent configurations. Formerly *Chess*. 58 figures. 102pp. (USO) 21210-6 Paperbound $1.25

LASKER'S MANUAL OF CHESS, Dr. Emanuel Lasker. Lasker was not only one of the five great World Champions, he was also one of the ablest expositors, theorists, and analysts. In many ways, his Manual, permeated with his philosophy of battle, filled with keen insights, is one of the greatest works ever written on chess. Filled with analyzed games by the great players. A single-volume library that will profit almost any chess player, beginner or master. 308 diagrams. xli x 349pp.

20640-8 Paperbound $2.75

THE MASTER BOOK OF MATHEMATICAL RECREATIONS, Fred Schuh. In opinion of many the finest work ever prepared on mathematical puzzles, stunts, recreations; exhaustively thorough explanations of mathematics involved, analysis of effects, citation of puzzles and games. Mathematics involved is elementary. Translated by F. Göbel. 194 figures. xxiv + 430pp. 22134-2 Paperbound $4.00

MATHEMATICS, MAGIC AND MYSTERY, Martin Gardner. Puzzle editor for Scientific American explains mathematics behind various mystifying tricks: card tricks, stage "mind reading," coin and match tricks, counting out games, geometric dissections, etc. Probability sets, theory of numbers clearly explained. Also provides more than 400 tricks, guaranteed to work, that you can do. 135 illustrations. xii + 176pp.

20335-2 Paperbound $2.00

"ESSENTIAL GRAMMAR" SERIES

All you really need to know about modern, colloquial grammar. Many educational shortcuts help you learn faster, understand better. Detailed cognate lists teach you to recognize similarities between English and foreign words and roots—make learning vocabulary easy and interesting. Excellent for independent study or as a supplement to record courses.

ESSENTIAL FRENCH GRAMMAR, Seymour Resnick. 2500-item cognate list. 159pp.
(EBE) 20419-7 Paperbound $1.50

ESSENTIAL GERMAN GRAMMAR, Guy Stern and Everett F. Bleiler. Unusual shortcuts on noun declension, word order, compound verbs. 124pp.
(EBE) 20422-7 Paperbound $1.25

ESSENTIAL ITALIAN GRAMMAR, Olga Ragusa. 111pp.
(EBE) 20779-X Paperbound $1.25

ESSENTIAL JAPANESE GRAMMAR, Everett F. Bleiler. In Romaji transcription; no characters needed. Japanese grammar is regular and simple. 156pp.
21027-8 Paperbound $1.50

ESSENTIAL PORTUGUESE GRAMMAR, Alexander da R. Prista. vi + 114pp.
21650-0 Paperbound $1.35

ESSENTIAL SPANISH GRAMMAR, Seymour Resnick. 2500 word cognate list. 115pp.
(EBE) 20780-3 Paperbound $1.25

ESSENTIAL ENGLISH GRAMMAR, Philip Gucker. Combines best features of modern, functional and traditional approaches. For refresher, class use, home study. x + 177pp.
21649-7 Paperbound $1.75

A PHRASE AND SENTENCE DICTIONARY OF SPOKEN SPANISH. Prepared for U. S. War Department by U. S. linguists. As above, unit is idiom, phrase or sentence rather than word. English-Spanish and Spanish-English sections contain modern equivalents of over 18,000 sentences. Introduction and appendix as above. iv + 513pp.
20495-2 Paperbound $3.50

A PHRASE AND SENTENCE DICTIONARY OF SPOKEN RUSSIAN. Dictionary prepared for U. S. War Department by U. S. linguists. Basic unit is not the word, but the idiom, phrase or sentence. English-Russian and Russian-English sections contain modern equivalents for over 30,000 phrases. Grammatical introduction covers phonetics, writing, syntax. Appendix of word lists for food, numbers, geographical names, etc. vi + 573 pp. 6⅛ x 9¼. 20496-0 Paperbound $5.50

CONVERSATIONAL CHINESE FOR BEGINNERS, Morris Swadesh. Phonetic system, beginner's course in Pai Hua Mandarin Chinese covering most important, most useful speech patterns. Emphasis on modern colloquial usage. Formerly *Chinese in Your Pocket*. xvi + 158pp.
21123-1 Paperbound $1.75

How to Know the Wild Flowers, Mrs. William Starr Dana. This is the classical book of American wildflowers (of the Eastern and Central United States), used by hundreds of thousands. Covers over 500 species, arranged in extremely easy to use color and season groups. Full descriptions, much plant lore. This Dover edition is the fullest ever compiled, with tables of nomenclature changes. 174 full-page plates by M. Satterlee. xii + 418pp. 20332-8 Paperbound $3.00

Our Plant Friends and Foes, William Atherton DuPuy. History, economic importance, essential botanical information and peculiarities of 25 common forms of plant life are provided in this book in an entertaining and charming style. Covers food plants (potatoes, apples, beans, wheat, almonds, bananas, etc.), flowers (lily, tulip, etc.), trees (pine, oak, elm, etc.), weeds, poisonous mushrooms and vines, gourds, citrus fruits, cotton, the cactus family, and much more. 108 illustrations. xiv + 290pp. 22272-1 Paperbound $2.50

How to Know the Ferns, Frances T. Parsons. Classic survey of Eastern and Central ferns, arranged according to clear, simple identification key. Excellent introduction to greatly neglected nature area. 57 illustrations and 42 plates. xvi + 215pp. 20740-4 Paperbound $2.00

Manual of the Trees of North America, Charles S. Sargent. America's foremost dendrologist provides the definitive coverage of North American trees and tree-like shrubs. 717 species fully described and illustrated: exact distribution, down to township; full botanical description; economic importance; description of subspecies and races; habitat, growth data; similar material. Necessary to every serious student of tree-life. Nomenclature revised to present. Over 100 locating keys. 783 illustrations. lii + 934pp. 20277-1, 20278-X Two volumes, Paperbound $7.00

Our Northern Shrubs, Harriet L. Keeler. Fine non-technical reference work identifying more than 225 important shrubs of Eastern and Central United States and Canada. Full text covering botanical description, habitat, plant lore, is paralleled with 205 full-page photographs of flowering or fruiting plants. Nomenclature revised by Edward G. Voss. One of few works concerned with shrubs. 205 plates, 35 drawings. xxviii + 521pp. 21989-5 Paperbound $3.75

The Mushroom Handbook, Louis C. C. Krieger. Still the best popular handbook: full descriptions of 259 species, cross references to another 200. Extremely thorough text enables you to identify, know all about any mushroom you are likely to meet in eastern and central U. S. A.: habitat, luminescence, poisonous qualities, use, folklore, etc. 32 color plates show over 50 mushrooms, also 126 other illustrations. Finding keys. vii + 560pp. 21861-9 Paperbound $4.50

Handbook of Birds of Eastern North America, Frank M. Chapman. Still much the best single-volume guide to the birds of Eastern and Central United States. Very full coverage of 675 species, with descriptions, life habits, distribution, similar data. All descriptions keyed to two-page color chart. With this single volume the average birdwatcher needs no other books. 1931 revised edition. 195 illustrations. xxxvi + 581pp. 21489-3 Paperbound $5.00

AMERICAN FOOD AND GAME FISHES, David S. Jordan and Barton W. Evermann. Definitive source of information, detailed and accurate enough to enable the sportsman and nature lover to identify conclusively some 1,000 species and sub-species of North American fish, sought for food or sport. Coverage of range, physiology, habits, life history, food value. Best methods of capture, interest to the angler, advice on bait, fly-fishing, etc. 338 drawings and photographs. l + 574pp. 6⅝ x 9⅜.

22196-2 Paperbound $5.00

THE FROG BOOK, Mary C. Dickerson. Complete with extensive finding keys, over 300 photographs, and an introduction to the general biology of frogs and toads, this is the classic non-technical study of Northeastern and Central species. 58 species; 290 photographs and 16 color plates. xvii + 253pp.

21973-9 Paperbound $4.00

THE MOTH BOOK: A GUIDE TO THE MOTHS OF NORTH AMERICA, William J. Holland. Classical study, eagerly sought after and used for the past 60 years. Clear identification manual to more than 2,000 different moths, largest manual in existence. General information about moths, capturing, mounting, classifying, etc., followed by species by species descriptions. 263 illustrations plus 48 color plates show almost every species, full size. 1968 edition, preface, nomenclature changes by A. E. Brower. xxiv + 479pp. of text. 6½ x 9¼.

21948-8 Paperbound $6.00

THE SEA-BEACH AT EBB-TIDE, Augusta Foote Arnold. Interested amateur can identify hundreds of marine plants and animals on coasts of North America; marine algae; seaweeds; squids; hermit crabs; horse shoe crabs; shrimps; corals; sea anemones; etc. Species descriptions cover: structure; food; reproductive cycle; size; shape; color; habitat; etc. Over 600 drawings. 85 plates. xii + 490pp.

21949-6 Paperbound $4.00

COMMON BIRD SONGS, Donald J. Borror. 33⅓ 12-inch record presents songs of 60 important birds of the eastern United States. A thorough, serious record which provides several examples for each bird, showing different types of song, individual variations, etc. Inestimable identification aid for birdwatcher. 32-page booklet gives text about birds and songs, with illustration for each bird.

21829-5 Record, book, album. Monaural. $3.50

FADS AND FALLACIES IN THE NAME OF SCIENCE, Martin Gardner. Fair, witty appraisal of cranks and quacks of science: Atlantis, Lemuria, hollow earth, flat earth, Velikovsky, orgone energy, Dianetics, flying saucers, Bridey Murphy, food fads, medical fads, perpetual motion, etc. Formerly "In the Name of Science." x + 363pp.

20394-8 Paperbound $3.00

HOAXES, Curtis D. MacDougall. Exhaustive, unbelievably rich account of great hoaxes: Locke's moon hoax, Shakespearean forgeries, sea serpents, Loch Ness monster, Cardiff giant, John Wilkes Booth's mummy, Disumbrationist school of art, dozens more; also journalism, psychology of hoaxing. 54 illustrations. xi + 338pp.

20465-0 Paperbound $3.50

THE PRINCIPLES OF PSYCHOLOGY, William James. The famous long course, complete and unabridged. Stream of thought, time perception, memory, experimental methods—these are only some of the concerns of a work that was years ahead of its time and still valid, interesting, useful. 94 figures. Total of xviii + 1391pp.
20381-6, 20382-4 Two volumes, Paperbound $9.00

THE STRANGE STORY OF THE QUANTUM, Banesh Hoffmann. Non-mathematical but thorough explanation of work of Planck, Einstein, Bohr, Pauli, de Broglie, Schrödinger, Heisenberg, Dirac, Feynman, etc. No technical background needed. "Of books attempting such an account, this is the best," Henry Margenau, Yale. 40-page "Postscript 1959." xii + 285pp.
20518-5 Paperbound $3.00

THE RISE OF THE NEW PHYSICS, A. d'Abro. Most thorough explanation in print of central core of mathematical physics, both classical and modern; from Newton to Dirac and Heisenberg. Both history and exposition; philosophy of science, causality, explanations of higher mathematics, analytical mechanics, electromagnetism, thermodynamics, phase rule, special and general relativity, matrices. No higher mathematics needed to follow exposition, though treatment is elementary to intermediate in level. Recommended to serious student who wishes verbal understanding. 97 illustrations. xvii + 982pp.
20003-5, 20004-3 Two volumes, Paperbound $10.00

GREAT IDEAS OF OPERATIONS RESEARCH, Jagjit Singh. Easily followed non-technical explanation of mathematical tools, aims, results: statistics, linear programming, game theory, queueing theory, Monte Carlo simulation, etc. Uses only elementary mathematics. Many case studies, several analyzed in detail. Clarity, breadth make this excellent for specialist in another field who wishes background. 41 figures. x + 228pp.
21886-4 Paperbound $2.50

GREAT IDEAS OF MODERN MATHEMATICS: THEIR NATURE AND USE, Jagjit Singh. Internationally famous expositor, winner of Unesco's Kalinga Award for science popularization explains verbally such topics as differential equations, matrices, groups, sets, transformations, mathematical logic and other important modern mathematics, as well as use in physics, astrophysics, and similar fields. Superb exposition for layman, scientist in other areas. viii + 312pp.
20587-8 Paperbound $2.75

GREAT IDEAS IN INFORMATION THEORY, LANGUAGE AND CYBERNETICS, Jagjit Singh. The analog and digital computers, how they work, how they are like and unlike the human brain, the men who developed them, their future applications, computer terminology. An essential book for today, even for readers with little math. Some mathematical demonstrations included for more advanced readers. 118 figures. Tables. ix + 338pp.
21694-2 Paperbound $2.50

CHANCE, LUCK AND STATISTICS, Horace C. Levinson. Non-mathematical presentation of fundamentals of probability theory and science of statistics and their applications. Games of chance, betting odds, misuse of statistics, normal and skew distributions, birth rates, stock speculation, insurance. Enlarged edition. Formerly "The Science of Chance." xiii + 357pp.
21007-3 Paperbound $2.50

PLANETS, STARS AND GALAXIES: DESCRIPTIVE ASTRONOMY FOR BEGINNERS, A. E. Fanning. Comprehensive introductory survey of astronomy: the sun, solar system, stars, galaxies, universe, cosmology; up-to-date, including quasars, radio stars, etc. Preface by Prof. Donald Menzel. 24pp. of photographs. 189pp. 5¼ x 8¼.
21680-2 Paperbound $2.50

TEACH YOURSELF CALCULUS, P. Abbott. With a good background in algebra and trig, you can teach yourself calculus with this book. Simple, straightforward introduction to functions of all kinds, integration, differentiation, series, etc. "Students who are beginning to study calculus method will derive great help from this book." Faraday House Journal. 308pp.
20683-1 Clothbound $2.50

TEACH YOURSELF TRIGONOMETRY, P. Abbott. Geometrical foundations, indices and logarithms, ratios, angles, circular measure, etc. are presented in this sound, easy-to-use text. Excellent for the beginner or as a brush up, this text carries the student through the solution of triangles. 204pp.
20682-3 Clothbound $2.00

BASIC MACHINES AND HOW THEY WORK, U. S. Bureau of Naval Personnel. Originally used in U.S. Naval training schools, this book clearly explains the operation of a progression of machines, from the simplest—lever, wheel and axle, inclined plane, wedge, screw—to the most complex—typewriter, internal combustion engine, computer mechanism. Utilizing an approach that requires only an elementary understanding of mathematics, these explanations build logically upon each other and are assisted by over 200 drawings and diagrams. Perfect as a technical school manual or as a self-teaching aid to the layman. 204 figures. Preface. Index. vii + 161pp. 6½ x 9¼.
21709-4 Paperbound $2.50

THE FRIENDLY STARS, Martha Evans Martin. Classic has taught naked-eye observation of stars, planets to hundreds of thousands, still not surpassed for charm, lucidity, adequacy. Completely updated by Professor Donald H. Menzel, Harvard Observatory. 25 illustrations. 16 x 30 chart. x + 147pp.
21099-5 Paperbound $2.00

MUSIC OF THE SPHERES: THE MATERIAL UNIVERSE FROM ATOM TO QUASAR, SIMPLY EXPLAINED, Guy Murchie. Extremely broad, brilliantly written popular account begins with the solar system and reaches to dividing line between matter and nonmatter; latest understandings presented with exceptional clarity. Volume One: Planets, stars, galaxies, cosmology, geology, celestial mechanics, latest astronomical discoveries; Volume Two: Matter, atoms, waves, radiation, relativity, chemical action, heat, nuclear energy, quantum theory, music, light, color, probability, antimatter, antigravity, and similar topics. 319 figures. 1967 (second) edition. Total of xx + 644pp.
21809-0, 21810-4 Two volumes, Paperbound $5.75

OLD-TIME SCHOOLS AND SCHOOL BOOKS, Clifton Johnson. Illustrations and rhymes from early primers, abundant quotations from early textbooks, many anecdotes of school life enliven this study of elementary schools from Puritans to middle 19th century. Introduction by Carl Withers. 234 illustrations. xxxiii + 381pp.
21031-6 Paperbound $4.00

THE PHILOSOPHY OF THE UPANISHADS, Paul Deussen. Clear, detailed statement of upanishadic system of thought, generally considered among best available. History of these works, full exposition of system emergent from them, parallel concepts in the West. Translated by A. S. Geden. xiv + 429pp.
21616-0 Paperbound $3.50

LANGUAGE, TRUTH AND LOGIC, Alfred J. Ayer. Famous, remarkably clear introduction to the Vienna and Cambridge schools of Logical Positivism; function of philosophy, elimination of metaphysical thought, nature of analysis, similar topics. "Wish I had written it myself," Bertrand Russell. 2nd, 1946 edition. 160pp.
20010-8 Paperbound $1.50

THE GUIDE FOR THE PERPLEXED, Moses Maimonides. Great classic of medieval Judaism, major attempt to reconcile revealed religion (Pentateuch, commentaries) and Aristotelian philosophy. Enormously important in all Western thought. Unabridged Friedländer translation. 50-page introduction. lix + 414pp.
(USO) 20351-4 Paperbound $4.50

OCCULT AND SUPERNATURAL PHENOMENA, D. H. Rawcliffe. Full, serious study of the most persistent delusions of mankind: crystal gazing, mediumistic trance, stigmata, lycanthropy, fire walking, dowsing, telepathy, ghosts, ESP, etc., and their relation to common forms of abnormal psychology. Formerly *Illusions and Delusions of the Supernatural and the Occult.* iii + 551pp.   20503-7 Paperbound $4.00

THE EGYPTIAN BOOK OF THE DEAD: THE PAPYRUS OF ANI, E. A. Wallis Budge. Full hieroglyphic text, interlinear transliteration of sounds, word for word translation, then smooth, connected translation; Theban recension. Basic work in Ancient Egyptian civilization; now even more significant than ever for historical importance, dilation of consciousness, etc. clvi + 377pp. 6½ x 9¼.
21866-X Paperbound $4.95

PSYCHOLOGY OF MUSIC, Carl E. Seashore. Basic, thorough survey of everything known about psychology of music up to 1940's: essential reading for psychologists, musicologists. Physical acoustics; auditory apparatus; relationship of physical sound to perceived sound; role of the mind in sorting, altering, suppressing, creating sound sensations; musical learning, testing for ability, absolute pitch, other topics. Records of Caruso, Menuhin analyzed. 88 figures. xix + 408pp.
21851-1 Paperbound $3.50

THE I CHING (THE BOOK OF CHANGES), translated by James Legge. Complete translated text plus appendices by Confucius, of perhaps the most penetrating divination book ever compiled. Indispensable to all study of early Oriental civilizations. 3 plates. xxiii + 448pp.   21062-6 Paperbound $3.50

THE UPANISHADS, translated by Max Müller. Twelve classical upanishads: Chandogya, Kena, Aitareya, Kaushitaki, Isa, Katha, Mundaka, Taittiriyaka, Brhadaranyaka, Svetasvatara, Prasna, Maitriyana. 160-page introduction, analysis by Prof. Müller. Total of 670pp. 20992-X, 20993-8 Two volumes, Paperbound $7.50

JIM WHITEWOLF: THE LIFE OF A KIOWA APACHE INDIAN, Charles S. Brant, editor. Spans transition between native life and acculturation period, 1880 on. Kiowa culture, personal life pattern, religion and the supernatural, the Ghost Dance, breakdown in the White Man's world, similar material. 1 map. xii + 144pp.
22015-X Paperbound $1.75

THE NATIVE TRIBES OF CENTRAL AUSTRALIA, Baldwin Spencer and F. J. Gillen. Basic book in anthropology, devoted to full coverage of the Arunta and Warramunga tribes; the source for knowledge about kinship systems, material and social culture, religion, etc. Still unsurpassed. 121 photographs, 89 drawings. xviii + 669pp.
21775-2 Paperbound $5.00

MALAY MAGIC, Walter W. Skeat. Classic (1900); still the definitive work on the folklore and popular religion of the Malay peninsula. Describes marriage rites, birth spirits and ceremonies, medicine, dances, games, war and weapons, etc. Extensive quotes from original sources, many magic charms translated into English. 35 illustrations. Preface by Charles Otto Blagden. xxiv + 685pp.
21760-4 Paperbound $4.00

HEAVENS ON EARTH: UTOPIAN COMMUNITIES IN AMERICA, 1680-1880, Mark Holloway. The finest nontechnical account of American utopias, from the early Woman in the Wilderness, Ephrata, Rappites to the enormous mid 19th-century efflorescence; Shakers, New Harmony, Equity Stores, Fourier's Phalanxes, Oneida, Amana, Fruitlands, etc. "Entertaining and very instructive." *Times Literary Supplement.* 15 illustrations. 246pp.
21593-8 Paperbound $2.00

LONDON LABOUR AND THE LONDON POOR, Henry Mayhew. Earliest (c. 1850) sociological study in English, describing myriad subcultures of London poor. Particularly remarkable for the thousands of pages of direct testimony taken from the lips of London prostitutes, thieves, beggars, street sellers, chimney-sweepers, street-musicians, "mudlarks," "pure-finders," rag-gatherers, "running-patterers," dock laborers, cab-men, and hundreds of others, quoted directly in this massive work. An extraordinarily vital picture of London emerges. 110 illustrations. Total of lxxvi + 1951pp. $6\frac{5}{8}$ x 10.
21934-8, 21935-6, 21936-4, 21937-2 Four volumes, Paperbound $16.00

HISTORY OF THE LATER ROMAN EMPIRE, J. B. Bury. Eloquent, detailed reconstruction of Western and Byzantine Roman Empire by a major historian, from the death of Theodosius I (395 A.D.) to the death of Justinian (565). Extensive quotations from contemporary sources; full coverage of important Roman and foreign figures of the time. xxxiv + 965pp. 20398-0, 20399-9 Two volumes, Paperbound $7.00

AN INTELLECTUAL AND CULTURAL HISTORY OF THE WESTERN WORLD, Harry Elmer Barnes. Monumental study, tracing the development of the accomplishments that make up human culture. Every aspect of man's achievement surveyed from its origins in the Paleolithic to the present day (1964); social structures, ideas, economic systems, art, literature, technology, mathematics, the sciences, medicine, religion, jurisprudence, etc. Evaluations of the contributions of scores of great men. 1964 edition, revised and edited by scholars in the many fields represented. Total of xxix + 1381pp.   21275-0, 21276-9, 21277-7 Three volumes, Paperbound $10.50

INCIDENTS OF TRAVEL IN YUCATAN, John L. Stephens. Classic (1843) exploration of jungles of Yucatan, looking for evidences of Maya civilization. Stephens found many ruins; comments on travel adventures, Mexican and Indian culture. 127 striking illustrations by F. Catherwood. Total of 669 pp.
20926-1, 20927-X Two volumes, Paperbound $5.50

INCIDENTS OF TRAVEL IN CENTRAL AMERICA, CHIAPAS, AND YUCATAN, John L. Stephens. An exciting travel journal and an important classic of archeology. Narrative relates his almost single-handed discovery of the Mayan culture, and exploration of the ruined cities of Copan, Palenque, Utatlan and others; the monuments they dug from the earth, the temples buried in the jungle, the customs of poverty-stricken Indians living a stone's throw from the ruined palaces. 115 drawings by F. Catherwood. Portrait of Stephens. xii + 812pp.
22404-X, 22405-8 Two volumes, Paperbound $6.00

A NEW VOYAGE ROUND THE WORLD, William Dampier. Late 17-century naturalist joined the pirates of the Spanish Main to gather information; remarkably vivid account of buccaneers, pirates; detailed, accurate account of botany, zoology, ethnography of lands visited. Probably the most important early English voyage, enormous implications for British exploration, trade, colonial policy. Also most interesting reading. Argonaut edition, introduction by Sir Albert Gray. New introduction by Percy Adams. 6 plates, 7 illustrations. xlvii + 376pp. 6½ x 9¼.
21900-3 Paperbound $3.00

INTERNATIONAL AIRLINE PHRASE BOOK IN SIX LANGUAGES, Joseph W. Bátor. Important phrases and sentences in English paralleled with French, German, Portuguese, Italian, Spanish equivalents, covering all possible airport-travel situations; created for airline personnel as well as tourist by Language Chief, Pan American Airlines. xiv + 204pp.
22017-6 Paperbound $2.25

STAGE COACH AND TAVERN DAYS, Alice Morse Earle. Detailed, lively account of the early days of taverns; their uses and importance in the social, political and military life; furnishings and decorations; locations; food and drink; tavern signs, etc. Second half covers every aspect of early travel; the roads, coaches, drivers, etc. Nostalgic, charming, packed with fascinating material. 157 illustrations, mostly photographs. xiv + 449pp.
22518-6 Paperbound $4.00

NORSE DISCOVERIES AND EXPLORATIONS IN NORTH AMERICA, Hjalmar R. Holand. The perplexing Kensington Stone, found in Minnesota at the end of the 19th century. Is it a record of a Scandinavian expedition to North America in the 14th century? Or is it one of the most successful hoaxes in history. A scientific detective investigation. Formerly *Westward from Vinland*. 31 photographs, 17 figures. x + 354pp.
22014-1 Paperbound $2.75

A BOOK OF OLD MAPS, compiled and edited by Emerson D. Fite and Archibald Freeman. 74 old maps offer an unusual survey of the discovery, settlement and growth of America down to the close of the Revolutionary war: maps showing Norse settlements in Greenland, the explorations of Columbus, Verrazano, Cabot, Champlain, Joliet, Drake, Hudson, etc., campaigns of Revolutionary war battles, and much more. Each map is accompanied by a brief historical essay. xvi + 299pp. 11 x 13¾.
22084-2 Paperbound $7.00

ADVENTURES OF AN AFRICAN SLAVER, Theodore Canot. Edited by Brantz Mayer. A detailed portrayal of slavery and the slave trade, 1820-1840. Canot, an established trader along the African coast, describes the slave economy of the African kingdoms, the treatment of captured negroes, the extensive journeys in the interior to gather slaves, slave revolts and their suppression, harems, bribes, and much more. Full and unabridged republication of 1854 edition. Introduction by Malcom Cowley. 16 illustrations. xvii + 448pp. 22456-2 Paperbound $3.50

MY BONDAGE AND MY FREEDOM, Frederick Douglass. Born and brought up in slavery, Douglass witnessed its horrors and experienced its cruelties, but went on to become one of the most outspoken forces in the American anti-slavery movement. Considered the best of his autobiographies, this book graphically describes the inhuman treatment of slaves, its effects on slave owners and slave families, and how Douglass's determination led him to a new life. Unaltered reprint of 1st (1855) edition. xxxii + 464pp. 22457-0 Paperbound $3.50

THE INDIANS' BOOK, recorded and edited by Natalie Curtis. Lore, music, narratives, dozens of drawings by Indians themselves from an authoritative and important survey of native culture among Plains, Southwestern, Lake and Pueblo Indians. Standard work in popular ethnomusicology. 149 songs in full notation. 23 drawings, 23 photos. xxxi + 584pp. 6⅝ x 9⅜. 21939-9 Paperbound $5.00

DICTIONARY OF AMERICAN PORTRAITS, edited by Hayward and Blanche Cirker. 4024 portraits of 4000 most important Americans, colonial days to 1905 (with a few important categories, like Presidents, to present). Pioneers, explorers, colonial figures, U. S. officials, politicians, writers, military and naval men, scientists, inventors, manufacturers, jurists, actors, historians, educators, notorious figures, Indian chiefs, etc. All authentic contemporary likenesses. The only work of its kind in existence; supplements all biographical sources for libraries. Indispensable to anyone working with American history. 8,000-item classified index, finding lists, other aids. xiv + 756pp. 9¼ x 12¾. 21823-6 Clothbound $30.00

TRITTON'S GUIDE TO BETTER WINE AND BEER MAKING FOR BEGINNERS, S. M. Tritton. All you need to know to make family-sized quantities of over 100 types of grape, fruit, herb and vegetable wines; as well as beers, mead, cider, etc. Complete recipes, advice as to equipment, procedures such as fermenting, bottling, and storing wines. Recipes given in British, U. S., and metric measures. Accompanying booklet lists sources in U. S. A. where ingredients may be bought, and additional information. 11 illustrations. 157pp. 5⅝ x 8⅛. 22090-7 **Paperbound $2.00**

GARDENING WITH HERBS FOR FLAVOR AND FRAGRANCE, Helen M. Fox. How to grow herbs in your own garden, how to use them in your cooking (over 55 recipes included), legends and myths associated with each species, uses in medicine, perfumes, etc.—these are elements of one of the few books written especially for American herb fanciers. Guides you step-by-step from soil preparation to harvesting and storage for each type of herb. 12 drawings by Louise Mansfield. xiv + 334pp. 22540-2 Paperbound $2.50

MATHEMATICAL PUZZLES FOR BEGINNERS AND ENTHUSIASTS, Geoffrey Mott-Smith. 189 puzzles from easy to difficult—involving arithmetic, logic, algebra, properties of digits, probability, etc.—for enjoyment and mental stimulus. Explanation of mathematical principles behind the puzzles. 135 illustrations. viii + 248pp.
20198-8 Paperbound $2.00

PAPER FOLDING FOR BEGINNERS, William D. Murray and Francis J. Rigney. Easiest book on the market, clearest instructions on making interesting, beautiful origami. Sail boats, cups, roosters, frogs that move legs, bonbon boxes, standing birds, etc. 40 projects; more than 275 diagrams and photographs. 94pp.
20713-7 Paperbound $1.00

TRICKS AND GAMES ON THE POOL TABLE, Fred Herrmann. 79 tricks and games—some solitaires, some for two or more players, some competitive games—to entertain you between formal games. Mystifying shots and throws, unusual caroms, tricks involving such props as cork, coins, a hat, etc. Formerly *Fun on the Pool Table.* 77 figures. 95pp.
21814-7 Paperbound $1.25

HAND SHADOWS TO BE THROWN UPON THE WALL: A SERIES OF NOVEL AND AMUSING FIGURES FORMED BY THE HAND, Henry Bursill. Delightful picturebook from great-grandfather's day shows how to make 18 different hand shadows: a bird that flies, duck that quacks, dog that wags his tail, camel, goose, deer, boy, turtle, etc. Only book of its sort. vi + 33pp. 6½ x 9¼.   21779-5 Paperbound $1.00

WHITTLING AND WOODCARVING, E. J. Tangerman. 18th printing of best book on market. "If you can cut a potato you can carve" toys and puzzles, chains, chessmen, caricatures, masks, frames, woodcut blocks, surface patterns, much more. Information on tools, woods, techniques. Also goes into serious wood sculpture from Middle Ages to present, East and West. 464 photos, figures. x + 293pp.
20965-2 Paperbound $2.50

HISTORY OF PHILOSOPHY, Julián Marias. Possibly the clearest, most easily followed, best planned, most useful one-volume history of philosophy on the market; neither skimpy nor overfull. Full details on system of every major philosopher and dozens of less important thinkers from pre-Socratics up to Existentialism and later. Strong on many European figures usually omitted. Has gone through dozens of editions in Europe. 1966 edition, translated by Stanley Appelbaum and Clarence Strowbridge. xviii + 505pp.
21739-6 Paperbound $3.50

YOGA: A SCIENTIFIC EVALUATION, Kovoor T. Behanan. Scientific but non-technical study of physiological results of yoga exercises; done under auspices of Yale U. Relations to Indian thought, to psychoanalysis, etc. 16 photos. xxiii + 270pp.
20505-3 Paperbound $2.50

Prices subject to change without notice.
Available at your book dealer or write for free catalogue to Dept. GI, Dover Publications, Inc., 180 Varick St., N. Y., N. Y. 10014. Dover publishes more than 150 books each year on science, elementary and advanced mathematics, biology, music, art, literary history, social sciences and other areas.